# Guide to Characteristics and Characterization of Semiconductor Surfaces

# Guide to Characteristics and Characterization of Semiconductor Surfaces

Jerzy Ruzyllo
Penn State University, USA

NEW JERSEY • LONDON • SINGAPORE • BEIJING • SHANGHAI • TAIPEI • CHENNAI

*Published by*

World Scientific Publishing Co. Pte. Ltd.

5 Toh Tuck Link, Singapore 596224

*USA office:* 27 Warren Street, Suite 401-402, Hackensack, NJ 07601

*UK office:* 57 Shelton Street, Covent Garden, London WC2H 9HE

**Library of Congress Cataloging-in-Publication Data**
Names: Rużyłło, Jerzy author
Title: Guide to characteristics and characterization of semiconductor surfaces /
Jerzy Ruzyllo, Penn State University, USA.
Description: Singapore ; Hackensack, NJ : World Scientific, [2025] |
Includes bibliographical references and index.
Identifiers: LCCN 2024048577 | ISBN 9789811254819 hardcover |
ISBN 9789811254826 ebook for institutions | ISBN 9789811254833 ebook for individuals
Subjects: LCSH: Semiconductors--Surfaces | Semiconductors--Characterization
Classification: LCC QC611.6.S9 R89 2025 | DDC 621.3815/2--dc23/eng/20250330
LC record available at https://lccn.loc.gov/2024048577

**British Library Cataloguing-in-Publication Data**
A catalogue record for this book is available from the British Library.



For any available supplementary material, please visit
https://www.worldscientific.com/worldscibooks/10.1142/12792#t=suppl

Desk Editors: Soundararajan Raghuraman/Steven Patt

Typeset by Stallion Press
Email: enquiries@stallionpress.com

*With Damon and Alina on my mind...*

# Preface

In semiconductor device technology, the characteristics and condition of the surfaces of semiconductor substrates, upon which electronic and photonic devices are formed, play a pivotal role in defining the performance and reliability of the final product. The purpose of this book is to give readers familiar with the fundamentals of semiconductor devices and process technology an opportunity to review topics related specifically to semiconductor surface characteristics, surface processing, and characterization as seen from the perspective of applied research geared toward device process technology rather than an in-depth discussion of surface science-related issues.

The topics considered in this book and the information provided herein are meant to expose the readers to challenges and solutions related to the characteristics and characterization of semiconductor surfaces with focus on techniques compatible with in-line process monitoring involving bare, as-processed semiconductor surfaces following cleaning and etching operations.

The contents of this contribution are based on the author's experiences in research and teaching in the field of semiconductor engineering including surface processing and characterization and are supported by results presented in research papers in this area published with his colleagues in academia, graduate students, and industrial collaborators.

This book is expected to address the needs of students pursuing advanced degrees in semiconductor electronics and photonics, materials engineering, and other related areas, as well as researchers, and professionals in the semiconductor industry who need to gain insights into practical aspects of semiconductor surface engineering, including surface characteristics, processing techniques, and characterization methods used in device process development, diagnostics, and monitoring. Although the discussion is based mostly on elemental silicon featuring a planar surface as a reference, selected aspects of surface processing in the case of germanium, silicon carbide and other compound semiconductor materials are also considered.

The volume starts with an introductory Chapter 1 in which the semiconductor surface is considered among other components of a material system functioning as a semiconductor device. Chapter 2 discusses the effect of the surface on the characteristics of semiconductor materials with recognition of the fact that in the cases where the architecture of the device departs from planarity, the role of the surface and the effect it has on its processing and device performance evolve accordingly. The follow-up Chapter 3 considers the interactions of semiconductor surfaces with external influences including ambient, electric field, light, as well as high energy electrons and ions, and temperature. The characteristics of the semiconductor surface defining its condition, such as surface energy, surface roughness, and surface charge are reviewed in Chapter 4. The goal of the subsequent Chapter 5 is to identify semiconductor devices in which surface effects play a performance-defining role including those featuring 3D surface geometry. The follow-up Chapter 6 surveys the methods of semiconductor surface processing, focused on cleaning and conditioning used in semiconductor device manufacturing.

Semiconductor surface characterization techniques, including analytical, optical, and electrical methods, are reviewed in Chapter 7. Among them of interest are methods geared specifically toward the characterization of electronic properties of semiconductor surfaces and near-surface region. This volume is concluded with Chapter 8, in which issues involved in semiconductor surface characterization, specifically in in-line process monitoring applications, are considered. The goal is to provide suggestions regarding the selection of methods which could potentially address needs around this topic.

*Jerzy Ruzyllo*
University Park, Pennsylvania

# Acknowledgements

Semiconductor surface engineering research experiences acquired by the author over the years were possible due to the interactions and close collaboration with many colleagues at Penn State University and earlier at Warsaw Institute of Technology. Moreover, it is with pleasure and gratitude that the author acknowledges Penn State graduate students, now accomplished professionals, working under his supervision over the years, who contributed greatly to the knowledge upon which the discussion in this book is based.

This *Guide* would never come to fruition if it were not for the collaboration with several professional colleagues and friends from academia and industry. During the very early years, it was Andrzej Jakubowski who was instrumental in getting me started with my research activities involving surface processing at the Warsaw University of Technology. Then, Nishizawa-sensei and colleagues at Tohoku University Ikuo Shiota and Nobuo Miyamoto helped me pursue surface characterization research. Later, in semiconductor surface research specifically, Rich Novak, Emil Kamieniecki, and Marc Heyns played an important role in shaping my interests which were further evolving through the collaboration and interactions with Bob Grant, Takeshi Hattori, Paul Mertens, Marc Meuris, Stefan De Gendt, Ara Philipossian, Francois Tardif, Adrien Danel, Allen Bowling, Paul Hammond, Paul Mumbauer, Rodney Ridley, Gary Dolny, Ismail Kashkoush, Mike Korwin-Pawlowski, Gilles Borsoni, Koichiro Saga, Pat Drummond and several others. Other than that, not much would have happened in my research activities without the support and patience of my wife Ewa.

# About the Author

**Jerzy Ruzyllo** is a Distinguished Professor Emeritus in the School of Electrical Engineering and Computer Science at Pennsylvania State University. He joined Penn State in 1984 after completing his education, obtaining a PhD degree in 1977, and serving on the faculty of Warsaw University of Technology in Poland where he is currently a Honorary Professor. Throughout his career, Dr. Ruzyllo was actively involved in research and teaching of semiconductor science and engineering. One of his research interests was focused on semiconductor surface processing and characterization. He was among the initiators of the series of *International Symposia on Semiconductor Cleaning Science and Technology* organized under the auspices of the Electrochemical Society since 1989 as well as *International Symposia on Ultra-Clean Processing of Semiconductor Surfaces* organized by IMEC since 1992.

Dr. Ruzyllo authored books titled *Semiconductor Glossary* in 2017 and *Guide to Semiconductor Engineering* in 2020. Dr. Ruzyllo is a Life Fellow of IEEE and a Fellow of the Electrochemical Society.

# Contents

# Chapter 1

# Surface as a Part of Semiconductor Device Material System

## 1.1 Introduction

Practical semiconductor devices are constructed based on complex material systems rather than on the stand-alone pieces of semiconductor materials. The discussion in this chapter defines the concept of a semiconductor material system and identifies its components which play the role in determining device performance, including substrates, surfaces, near-surface regions, and interfaces. It also considers thin films and contacts as integral parts of the material system comprising semiconductor devices, both discreet and integrated.

While all the above are considered in this chapter, the focus of the discussion in this volume will be solely on the broad range of issues related to characteristics and characterization of semiconductor surfaces.

## 1.2 Semiconductor Device Material System

To fabricate functional semiconductor devices (Pierret, 1996; Sze and Ng, 2006), semiconductor materials need to be incorporated into a material system comprising solids featuring diverse chemical compositions and crystallographic structures. The way they are integrated into the device as well as the selection of materials used are determined by the function any given device is designed to perform.

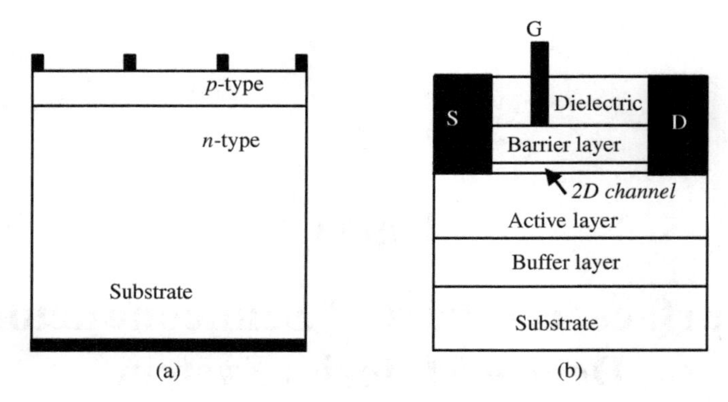


**Fig. 1.1.** Schematic cross-sections of two different types of semiconductor device material systems representing (a) *p–n* junction solar cell and (b) high-electron-mobility transistor (HEMT).

To illustrate how different in terms of complexity semiconductor devices can be, Fig. 1.1 shows simplified schematic cross-sections of a conventional *p–n* junction solar cell (Fig. 1.1(a)), the purpose of which is to convert light into an electric signal, and cross-section of the high-electron-mobility transistor (HEMT) constructed using compound semiconductors (Fig. 1.1(b)) and used in circuitry controlling communication in the GHz frequency regime.

As seen in Fig. 1.1, the material system forming HEMT is significantly more complex than in the case of solar cells. Yet, both solar cells and the HEMT involve the same key components although the manufacturing process in the former case is simpler than in the latter. Accordingly, procedures used in the fabrication of devices such as those shown in Figs. 1.1(a) and 1.1(b) vary significantly.

Accepting the notion that any functional semiconductor device is constructed using fundamentally the same building blocks, Fig. 1.2 identifies key elements of the material system comprising a semiconductor device using metal–oxide–semiconductor field-effect transistor (MOSFET) as an example (Grove, 1967; Pierret, 1983).

The starting element in any semiconductor device fabrication sequence is a substrate upon which the remaining parts of the device are formed. Depending on the application, substrates can be electrically conductive or not, and optically transparent or not. Of interest in this discussion are single-crystal semiconductor wafers, for instance silicon acting as substrates. The surface of such substrates is an adequate representation of the issues concerned with semiconductor surface characteristics and characterization on which the discussion in this book is focused.

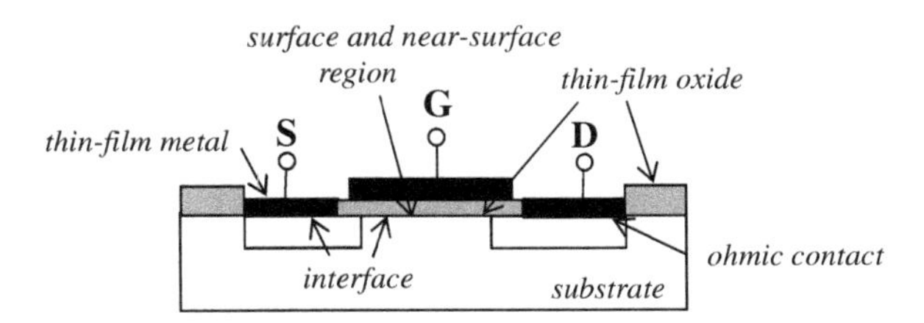


**Fig. 1.2.** Cross-section of the MOSFET used here to identify the key components of a typical semiconductor device material system.

The elements of the material system identified in Fig. 1.2 constitute integral parts of essentially any semiconductor device (Irene, 2008, Ruzyllo, 2020). The difference in terms of materials used and structural complexity of the material system involved depends on the function of the device and can differ significantly, as shown in Fig. 1.1. In the following sections, components of the material system exemplified in Fig. 1.2 are briefly considered.

## 1.3 Substrate

Semiconductor device fabrication sequence starts with a substrate upon which other parts of the device are formed. In mainstream applications, the substrate is a semiconductor material acting as an active part of the device while at the same time providing mechanical support for the device structure formed upon it. The MOSFET shown in Fig. 1.2 in which a channel is induced in the substrate between source and drain regions is an example of a case where the substrate participates in device operation. Alternatively, substrates made from materials which are not semiconductors, and which can be electrically conductive or isolating, optically transparent or opaque, flexible or rigid, can be used to provide mechanical support for devices formed on their surfaces.

**Semiconductor substrates.** In the typical electronic and photonic device applications which use semiconductor substrates, the term "substrate" is commonly associated with a round-shaped wafer of single-crystal semiconductor material. Depending on the material and needs of the device process technology, single-crystal semiconductor wafers may vary in diameter from less than 20 mm to 450 mm with thickness varying from less than about 0.1 mm to about 1.0 mm depending on the size of the wafer and its use. In some applications, where the cost of the substrate material is a major contributor to the overall cost, as it is in the case of solar cell manufacturing for instance, polycrystalline substrate wafers can be used instead of single-crystal substrates.

Considering the characteristics of semiconductor materials used in the manufacturing of functional devices, there is a need to distinguish between surface and near-surface region, and the bulk of the substrate as physical characteristics of these parts of semiconductor substrate vary significantly. In some cases, the properties of the bulk of the semiconductor play a role in determining the performance of the device. In others, the characteristics of the surface, near-surface region, and interfaces with other materials forming the device affect the working of the device the most, regardless of whether the device is electronic or photonic (Lukasiak *et al.*, 2001). The role of the bulk and near-surface region depends on the mode of operation and structure of the final device as well as whether it is processed into the semiconductor wafer substrate or is based on thin-film technology in which the substrate is not involved in device operation.

Despite mechanical cohesiveness and continuity of crystallographic structure, the near-surface region and bulk of the substrate wafer feature distinctly different electronic characteristics (Fig. 1.3).

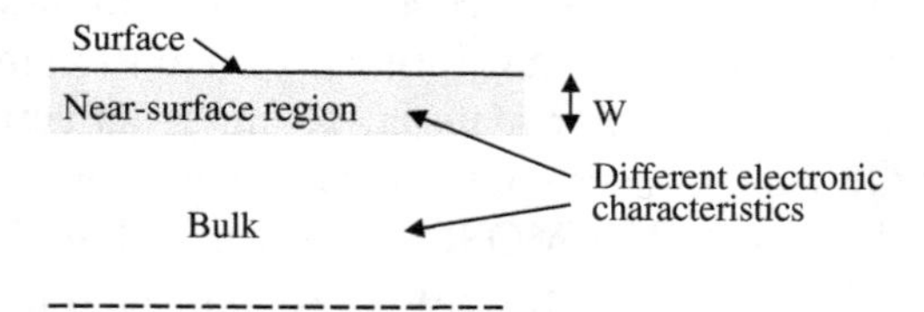


**Fig. 1.3.** Illustration of the surface, near-surface region, and the bulk of the semiconductor substrate wafer.

In other words, surface characteristics within the space charge region featuring depth $W$, which reflects the presence of an electric field caused by electrically active centers in the near-surface region, depart significantly from the same characteristics, or the same parameters, in the bulk of the semiconductor. For instance, due to electrically charged centers resulting from the defective lattice in the near-surface region of semiconductor wafers, increased recombination rate lowering carriers' lifetime $\tau$ and increased scattering of charge carriers lowering carriers' mobility $\mu$ in the near-surface region as compared to the values of the same parameters in the bulk of the semiconductor are observed. This effect has a major impact on the semiconductor devices' operation which depends on the surface and interface-related phenomena such as, for instance, metal–oxide–semiconductor field effect transistor (MOSFET) shown in Fig. 1.2.

Bulk single-crystal material is typically obtained in the form of an ingot using methods of single-crystal formation such as the Chochralski (CZ) growth method. Ingots are sliced into wafers along the designated crystallographic planes to establish the desired surface orientation. The wafers are then subject to operations carried out to establish the adequate finish of the wafer surface. More detailed considerations regarding surface finishing steps at this stage of the substrate wafer processing can be found in Chapter 2.

In some cases, surface finishing operations cannot address all the issues related to the preparation of high-quality surface and near-surface region. For instance, in the case of particularly mechanically hard semiconductor materials, such as silicon carbide, SiC, additional steps need to be taken to bring the surface of the wafer to the condition commensurate with the needs of the high-performance device fabrication process. Furthermore, even wafers with the surface polished to a mirror-like condition may require additional steps to minimize the effect of near-surface contamination introduced by polishing operations such as dopant deactivation observed in *p*-type silicon wafers (Roman, Staffa, *et al.*, 1998).

Dedicated processes referred to as wafer engineering are used to upgrade the condition of the substrate wafer to device-ready quality. Among various wafer engineering approaches, of interest in this discussion are processes which are implemented for the purpose of improving the characteristics of the substrate surface and near-surface region. Surface and near-surface region-related wafer engineering approaches include denuded zone formation and epitaxial extension (Fig. 1.4).

The process of denuded zone formation is used to form a few micrometers thick region immediately adjacent to the surface which is free from the electrically active defects and undesired contaminating elements which were incorporated into the wafer during processing (Fig. 1.4(a)). This is accomplished by employing the process of either intrinsic or extrinsic gettering with both enforcing the relocation of selected contaminants and defects away from the surface toward the bulk of the wafer. The process of intrinsic gettering involves elaborate sequences of elevated temperature treatments, while the process of extrinsic gettering uses external physical interactions inducing stress at the back surface of the wafer. At the elevated temperature, induced stress promotes the motion of certain elements toward the back surface of the wafer away from the front surface where the denuded zone is formed.

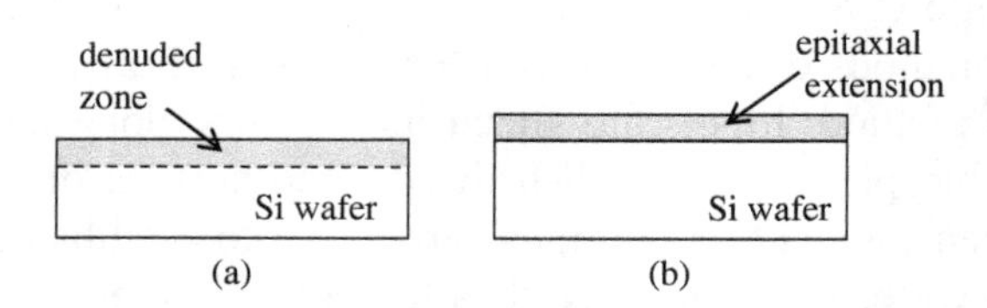


**Fig. 1.4.** Silicon substrate wafers featuring (a) denuded zone and (b) epitaxial extension.

The processes involved in wafer manufacturing may leave the surface of the substrate wafer physically damaged beyond the improvement of its condition by means of gettering and thermal annealing. In such cases, epitaxial extension, adding a layer of single-crystal material to the substrate, and thus creating a higher quality surface away from the original one, is a solution (Fig. 1.4(b)). Epitaxial extension needs to be sufficiently thick so as not to reproduce the defects of the original surface in the newly formed surface region. In the case of mechanically hard silicon carbide (SiC) wafers, in which case complete damage removal by polishing is not possible, epitaxial layer thickness needs to be in the range of tens of micrometers. In the case of high-quality Si substrate wafers where epitaxial extension is needed because of the requirements imposed by device architecture, the thickness of the epitaxial extension is on the order of single micrometers.

**Non-semiconductor substrates.** There are several semiconductor device applications in which substrates in the form of rigid, crystalline, non-transparent, electrically conductive semiconductor wafers are not needed or not desired. In such cases, insulating substrates are employed to provide mechanical support for semiconductor devices built on their surfaces using broadly understood thin-film technology.

One such substrate material is a single-crystal hexagonal form of aluminum oxide $Al_2O_3$, known as sapphire, the properties of which are highly conducive to the needs of several key electronic and photonic semiconductor devices. Among other insulators, sapphire distinguishes itself with superior resistance to temperature (melting point of 2300°C) as well as resistance to aggressive chemistries and high energy radiation. An added advantage is the high transparency of sapphire (over 80%) to light with wavelengths ranging from about 0.3 $\mu$m to about 4 $\mu$m.

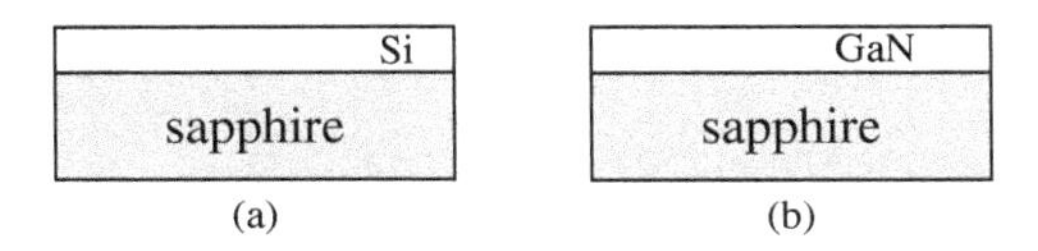


**Fig. 1.5.** Examples of sapphire used as a substrate in semiconductor devices: (a) in silicon-on-sapphire (SOS) configuration and (b) in gallium nitride (GaN) device fabrication.

In silicon technology, sapphire is applied as a substrate in silicon-on-sapphire (SOS) material systems (Fig. 1.5(a)) used primarily in the manufacturing of high-frequency analog integrated circuits. In III–V device technology, sapphire is used as a substrate for gallium nitride (GaN) in the fabrication of blue light emitting diodes (LEDs) and power electronic devices (Fig. 1.5(b)). In both cases, a mismatch of crystal lattices between sapphire and either of these two semiconductor materials is worked around using elaborate treatments preceding epitaxial deposition of thin semiconductor films. A brief overview of sapphire surface treatments demonstrating that sapphire surface processing in many aspects is similar to the procedures involved in semiconductor surface processing can be found in later parts of this book.

Among non-crystalline substrates used in thin-film semiconductor device manufacturing, glass is the most common. In applications in which transparency of the substrate is required, but at the same its surface needs to be electrically conductive, glass slides are covered with a film of material that is acting as a conductor which at the same time is transparent to visible light. Indium-tin-oxide, ITO in short, is the most common transparent conductor used for this purpose and ITO-covered glass substrates are readily available commercially. If flexible semiconductor electronic circuits, or light-emitting diodes, or solar cells are required, bendable and rollable plastic films, most often polyimides, are used as the substrates upon which thin-film semiconductor devices are formed.

In this volume, the discussion of the effect of surface characteristics on device performance is focused on single-crystal semiconductor substrates, primarily silicon. It needs to be emphasized, however, that many of the challenges associated with the characteristics and characterization of the surfaces in semiconductor device engineering are not specific to the single-crystal substrates and can be extended to other substrates such as those considered above.

## 1.4 Surface and Near-Surface Region

As the characteristics of semiconductor surfaces are the focus of the discussion in this book, the goal of this section is only to introduce the concept of semiconductor surfaces and briefly consider the role they play as an integral part of the semiconductor device (Fig. 1.2). The discussion continues in the later parts of this guide, with more detailed considerations of the characteristics of semiconductor surfaces and the role they play in semiconductor device engineering.

Parts of the semiconductor device material system, identified in Fig. 1.2 as surface and near-surface region, play an important role in determining the characteristics of semiconductor devices of which it is a part. However, such recognition of the role of the surface is valid only in the cases of materials featuring predictable, reproducible, and well-controlled bulk characteristics. In other words, in the case of semiconductor materials featuring inferior bulk characteristics such as disordered crystal structure, high density of defects, and poorly controlled chemical composition, the role of the surface in determining the performance of an overall material system is secondary to the effect of the bulk characteristics of semiconductor materials used.

For this reason, in the discussion of semiconductor surfaces in this volume, the assumption is made that the substrate comprises a high-quality single-crystal material. Only then the role of the surfaces in defining device characteristics can be adequately separated from other factors affecting its performance.

In general terms, the surface is an exterior face of the solid and represents the two-dimensional termination of the fundamental characteristics displayed by three-dimensionally distributed atoms in the bulk.

As indicated earlier in Fig. 1.3, the surface and near-surface region feature properties distinctly different than in the bulk of the same wafer. Figure 1.6 shows an attempt to graphically represent in a simplified fashion the nature of these differences showing unsaturated, broken bonds, known as dangling bonds, which are represented by chemically and electrically active surface states prone to interactions with the ambient.

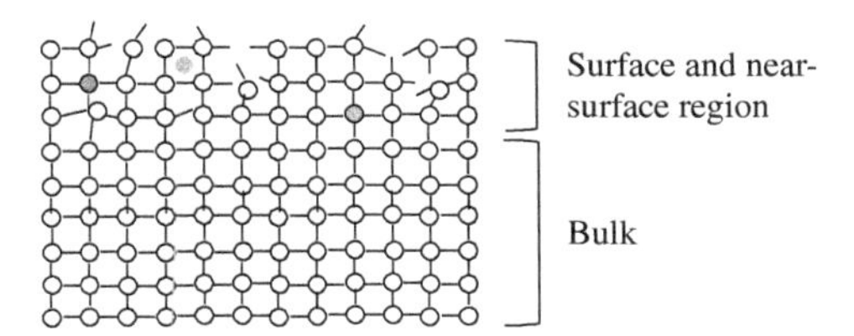


**Fig. 1.6.** Graphical illustration of the state of the near-surface region of semiconductor wafer expanding on the schematic representation of this region of semiconductor substrate in Fig. 1.3.

Unless neutralized, dangling bonds may cause changes in the distribution of electric charge in the sub-surface region of semiconductors. Furthermore, the surface region features missing atoms (vacancies), interstitially located host (interstitial atoms) and alien atoms (contaminants), as well as products of the wafer surface interactions with the process ambient. As shown in Fig. 1.6, the impact of the surface on the properties of a solid is not limited to the single-atom plane at the surface but extends to the near-surface region.

Considering the above comments and reiterating points made earlier, it is evident that the electronic properties of the near-surface region of semiconductor material are different from the same properties in its bulk. The effect of this difference, as seen from the perspective of surface characterization and semiconductor device engineering, is further considered in the later parts of this book.

## 1.5 Interface

The interfaces between materials featuring different structures and chemical compositions are inherent to any material system forming semiconductor devices and, in many instances, define the performance of such devices. An interface is understood here as a region featuring finite thickness and located in between two materials in physical contact. Its properties depend on the materials involved and the way the material system was created and chemically interweaved. In practical devices such as MOSFET shown in Fig. 1.2, there are two interfaces that influence device operation. The first one is an interface between the semiconductor substrate and gate oxide, and the other the interface between conductors, commonly metals or metal alloys, forming drain and source contacts, and semiconductors.

An interface is essentially a region that is needed to accommodate structural transition between two materials featuring either different crystal structures or compositional transition between materials featuring different chemical compositions. In practical semiconductor material systems acting as functional devices, typically both structural and chemical transitions occur at the same time.

As an example, Fig. 1.7(a) shows transmission electron microscopy (TEM) micrograph visualizing a cross-section of the MOSFET gate structure where amorphous high-*k* gate dielectric $HfO_2$ is interfaced with single-crystal silicon (Chang *et al.*, 2004). As the micrograph shown demonstrates, the interface in the case considered includes the distorted near-surface region of the silicon substrate which was not reconstructed either prior to or during high-*k* dielectric deposition. This is not the case when gate dielectric is a thermally grown silicon dioxide, $SiO_2$ (Fig. 1.7(b)). During the thermal oxidation process, atoms of silicon in the damaged top atomic layers are oxidized, a near-surface region is reconstructed, and a less defective interface than in the case of deposited $HfO_2$ films results. Comparison of TEM micrographs in Fig. 1.7 testifies to this effect.

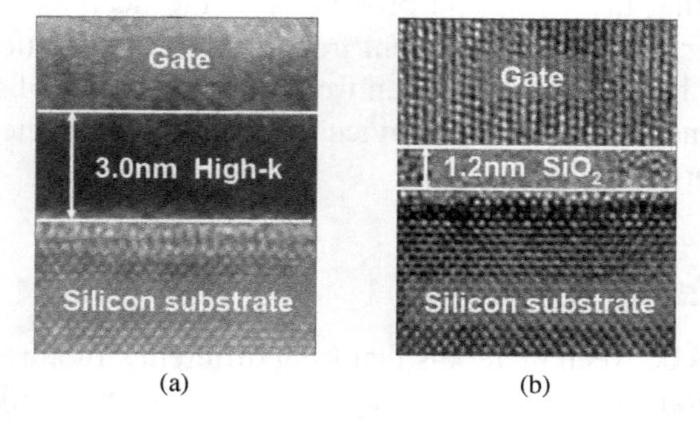


**Fig. 1.7.** Transmission electron microscopy (TEM) micrographs showing the distorted interface region between (a) high-*k* dielectric $HfO_2$ and single-crystal silicon and (b) thermally grown $SiO_2$ and single-crystal silicon.

Regardless of its configuration, the interface represents a discontinuity of electrical, optical, mechanical, and thermal properties of the material system including semiconductor devices and basically can be seen as a planar defect. Whatever the nature of the difference between two materials in contact, whether concerned with electrical or optical characteristics, the difference is accommodated within the interface transition region featuring finite thickness.

An issue of importance to the processes establishing the final characteristics of an interface is concerned with the condition of the surface of the semiconductor substrate upon which the material is deposited and the interface is formed. Processes preparing semiconductor surfaces prior to thin film deposition are considered in Chapter 6.

In addition to adequate surface preparation, processes establishing the final characteristics of an interface often involve annealing operations carried out at the elevated temperature in the gaseous ambient featuring the desired composition after film deposition. Under properly selected conditions, annealing processes may improve the characteristics of the interface in a variety of ways and as such are often an integral part of the interface formation process.

## 1.6 Thin Films

Thin films, understood here as the films of solids, regardless of their chemical composition, not exceeding in thickness of about 1 $\mu$m, are an integral part of any material system formed to act as an electronic or photonic semiconductor device. Films of dielectrics and metals in Fig. 1.2 are examples of where and how thin films are being used in the semiconductor device structure.

The concept of thin films is somewhat elusive considering the dependence of thin-film parameters on its thickness. The universally valid number defining thickness below which thin films rather than bulk properties define the characteristics of the material cannot be given because the thickness at which the transition from the bulk-dominated properties to thin-film properties occurs is different for different materials. In addition, it depends on the crystallographic structure of the material, its purity, and defect density, all of which are controlled by the thin-film deposition technique used (Ruzyllo, 2020).

In general, characteristics such as resistivity of the thin-film material depart from the bulk value as the thickness of the film is decreased into the ultra-thin regime. This relationship is illustrated in Fig. 1.8 where the resistivity of the semiconductor displays bulk characteristics as long as the continued reduction of the film thickness does not alter electrons' motion in any of the three directions. At a certain thickness, resistivity starts increasing as the increasing scattering of strongly 2D-confined electrons alters their flow. This is a point at which the material no longer displays bulk properties and assumes the properties of a thin film.

At the further reduced thickness, typically in the range of single atoms, the material transitions into the state of nano-confinement in which laws of classical physics no longer apply and quantum phenomena begin to dominate its physical properties. In this state, electrical resistivity has no effect on electrons flowing in the direction parallel to the surface as the scattering-free ballistic transport controls the motion of charge carriers in a two-dimensional (2D) material, which is a condition characteristic of electrons' behavior in a 2D material system known as quantum well (Fig. 1.8).

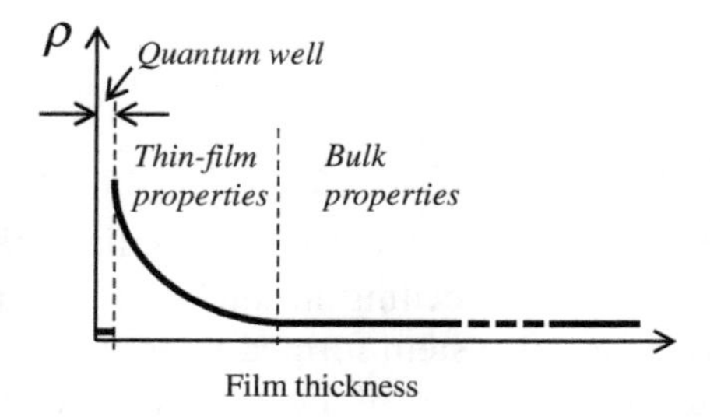


**Fig. 1.8.** Simplified qualitative illustration of the changes in semiconductor material resistivity $\rho$ as its thickness is decreasing.

The discussion of the thin-film semiconductors science and engineering, which for all practical purposes is a self-contained part of broadly understood semiconductor science and engineering (Irene, 2008; Lüth, 2015; Ohring, 2002), is beyond the scope of the discussion in this volume. An exception is the processing of the surface of the substrate upon which thin-film material is being deposited, which is considered in later parts of this volume.

## 1.7 Ohmic Contacts

The integral components of any semiconductor material system processed into a functional device are ohmic contacts through which electrical communication between the device and external electronic circuitry is established. The term "ohmic" is used in reference to contacts between conductive material, typically between metal and semiconductor, which feature very low, independent of applied voltage series resistance $R_s$. As a result, the current–voltage (*I–V*) characteristic of ohmic contact is linear, symmetric, and features a high slope, reflecting very low series resistance of the contact ($R_s \rightarrow 0$), as shown in Fig. 1.9.

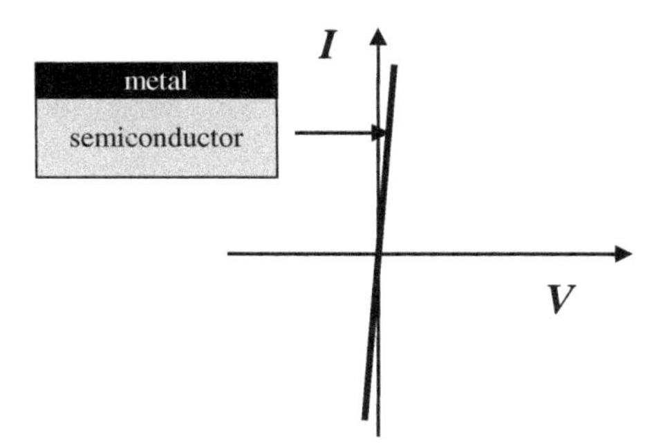


**Fig. 1.9.** Current–voltage (*I–V*) characteristic of an ohmic contact.

As an example of applications of ohmic contacts, Fig. 1.2 shows the source and drain ohmic contacts in the conventional MOS field effect transistor which need to feature *I–V* characteristic, shown in Fig. 1.9. The same ohmic contact-related considerations apply to any semiconductor device as in one way or another, all semiconductor devices need to be equipped with contacts assuring undisturbed flow of electric current in and out of the device.

In general, in the selection of materials forming ohmic contacts with any given semiconductor, as well as in the processing of such contacts, two factors need to be considered. First, the metal forming contact with the semiconductor must be selected such that there is no potential barrier formed at the interface, or the potential barrier is so thin that the electrons constituting the current can readily tunnel through it. Second, ohmic contact technology involves the preparation of the semiconductor surface prior to metal deposition.

The second observation makes considerations of the ohmic contact technology of interest in the discussion in this book. This is because in order to minimize series resistance $R_s$ of the contact, it is imperative that the presence of any interfacial film between metal and semiconductor, such as spontaneously grown oxide, is either prevented or oxide is removed immediately prior to metal deposition. The discussion related to this aspect of semiconductor device engineering is continued in the later parts of this book.

# Chapter 2

# Effect of Surface on Characteristics of Semiconductor Materials

## 2.1 Introduction

To the extent more pronounced in semiconductors than in the case of conductors or insulators, near-surface properties of materials vary from those in their bulk. The goal of the discussion in this chapter is to establish an understanding of the way the formation of the surface by the abrupt termination of the cohesive structure of semiconductor materials is altering its key properties.

In the first section of this chapter, processes involved in the formation of the surface of semiconductor materials used as substrates in device manufacturing processes are briefly reviewed. Subsequently, alterations of interatomic bonds, energy bands, as well as crystal structure and chemical composition in the near-surface region are considered. Then, the effect of the surface on selected electrical characteristics of semiconductor materials is discussed. Finally, the impact of the non-planarity of the surface and geometrical confinement of the semiconductor sample on the way the surface modifies its characteristics is considered.

Overall, the discussion in this chapter is meant to serve as an introduction to the considerations of various surface characteristics, processing methods, and characterization techniques discussed later in this volume. It is focused on single-crystal semiconductors with silicon being a semiconductor material of reference.

## 2.2 Processing of Semiconductor Wafer and Its Surface

As it is generally understood, the surface is the outermost layer of a physical object which represents abrupt discontinuity of its structure. It is also a part which interacts with an ambient or other materials with which it remains in direct contact. What needs to be emphasized at this point is that the formation of the surface is not a spontaneous process and the energy from an outside source is needed to create the surface. In nature, energy comes from various sources and often produces multi-surface objects featuring a broad range of shapes, crystallographic structures, and chemical compositions.

In contrast to the processes occurring in nature, materials used in engineering endeavors are man-made, feature controlled chemical composition and structure, and come in predetermined shapes and sizes. Depending on the manufacturing goals, the finish of the surfaces of such materials may not affect in any major way the performance of the final product, or just the opposite, the finish of the surface of the starting material may have a direct impact on the performance of the final product. Semiconductor device manufacturing not only belongs to the latter but also is undoubtedly among the most demanding in terms of the quality of the surfaces of semiconductor materials processed toward functional devices, such as diodes, discreet transistors, or integrated circuits.

In mainstream semiconductor device manufacturing, the starting material is in the form of a single-crystal wafer featuring a finely processed top surface. In order to better understand the nature of the technical challenges involved, procedures used in the fabrication of such wafers are briefly reviewed in this section. It needs to be remembered that in the case of some semiconductor materials, deficiencies of the surface finish cannot be eliminated entirely, and thus, their effect cannot be minimized without subjecting the wafer to additional treatments such as those discussed in Section 1.2 (Fig. 1.4).

The following discussion is focused on the processing of the surfaces of single-crystal silicon wafers, the fabrication of which is a good representation of the semiconductor surface processing at the wafer manufacturing stage.

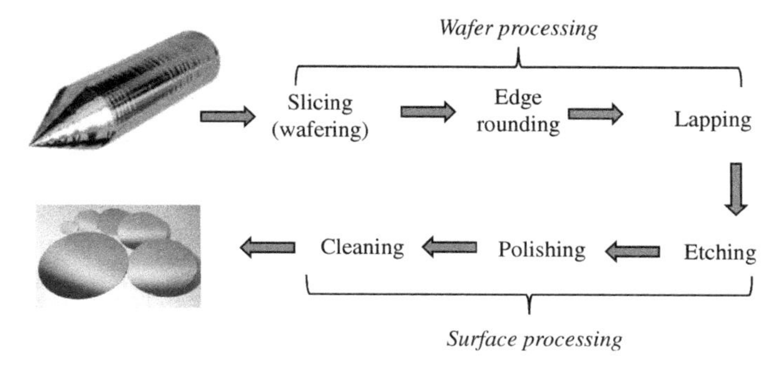


**Fig. 2.1.** From the bulk ingot to the wafer used to fabricate semiconductor devices through wafer processing and surface processing stages.

The bulk single-crystal silicon is typically obtained in the form of an ingot using methods, such as the Czochralski (CZ) crystal growth (Ruzyllo, 2017). As illustrated in Fig. 2.1, the very first step in the wafer fabrication sequence is an operation known as wafering in the course of which an ingot of single-crystal silicon is cut into wafers using high-precision inner diameter diamond-blade saw or multiple-wire saws.

In either case, the slicing procedure must precisely establish the desired crystallographic orientation of the surface of the resulting wafer. As it will be pointed out later in this chapter and will be again alluded to on several occasions later in this book, the crystallographic orientation of the surface of the single-crystal material impacts its basic characteristics as well as affects the way it interacts with other materials with which it forms an interface.

Following wafering procedures, each resulting wafer is subjected to a series of operations establishing its final geometry, and, most importantly, bringing its top surface after the relatively rough wafering process to the mirror-like smoothness. The sequence of operations preparing the surface of the wafer for the manufacture of semiconductor devices includes lapping and polishing performed on both sides using mechanical means and properly formulated slurry (Fig. 2.1). Lapping removes residual physical damage and defines the final geometrical characteristics of the wafer including its thickness, parallelism of front and back surfaces, and flatness. The goal of the chemical–mechanical polishing is to establish a mirror-finished surface of one or both sides of the wafer depending on the needs of the subsequent device manufacturing procedures (Philipossian and Mustapha, 2003). The undesired side effect of the polishing operation is that it may alter the chemical composition of the near-surface region of the wafer which can be penetrated by the elements contained in the polishing slurry, altering its electrical characteristics (Roman, Staffa, *et al.*, 1998).

A thorough cleaning, rinsing, and drying operations complete the wafer manufacturing sequence and preparation of its surface (Fig. 2.1). The importance of this sequence results from the fact that by being the very final stage in the wafer fabrication procedures, it establishes the chemical composition of the wafer surface (see a review of the wafer cleaning processes employed in semiconductor device applications in Chapter 6).

As a result of the surface processing operations, the top surface of the wafer and its near-surface region feature superior characteristics, including electronic properties, compared to the wafer sliced out of the single-crystal ingot (Fig. 2.1). To illustrate the effect of surface processing, Fig. 2.2 compares near-surface carrier mobility and carrier recombination lifetime for the just-sawed silicon wafers and the wafers featuring a top surface subjected to chemical polishing and cleaning (Drummond, Kshirsagar, *et al.*, 2011). Clearly, improved electronic properties in the latter case testify to the importance of surface processing in the manufacture of wafers used in the fabrication of semiconductor devices.

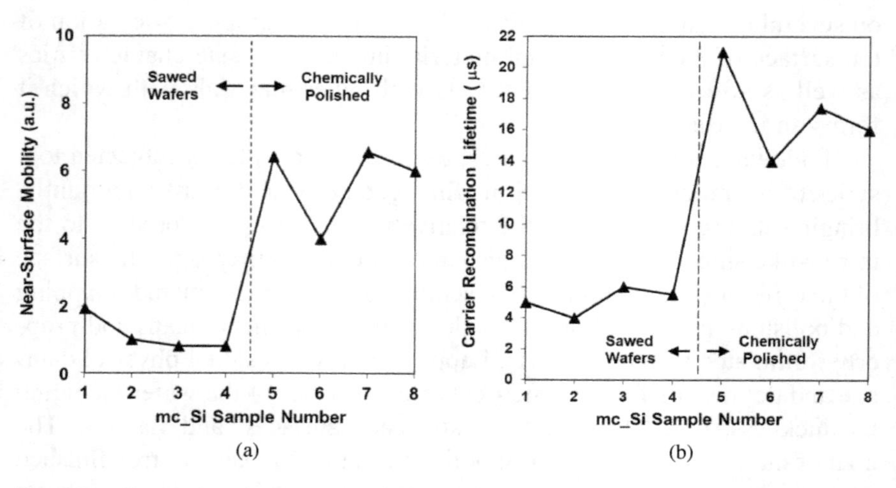


**Fig. 2.2.** Comparison between sawed Si wafers and chemically polished wafers in terms of near-surface (a) carrier mobility and (b) carrier recombination lifetime.

In the case of semiconductor materials featuring hardness precluding the elimination of the surface defects by means of operations outlined in Fig. 2.1, additional measures assuring adequate conditions of the surface must be taken. Silicon carbide (SiC) is an example of a single-crystal semiconductor material which because of its mechanical hardness is difficult to process into wafers featuring mirror-like surface finish. Consequently, the epitaxial extension discussed in Chapter 1 (Fig. 1.4) needs to be implemented to obtain the device-grade top surface of the SiC wafer.

## 2.3 Interatomic Bonds and Energy Bands at the Surface

The nature of bonds responsible for the cohesion of solids determines their fundamental properties including electrical conductivity discussed later in this chapter. Figure 2.3 shows silicon atoms of which valence electrons form covalent bonds responsible for the cohesion of materials.

This order is discontinued at the surface where the bonds are broken and remaining unsaturated are commonly referred to as dangling. Due to the missing electrons, the broken bonds on the surface are positively charged. This is causing the characteristics of the surface and near-surface region of single-crystal semiconductor materials to differ from those away from the surface.

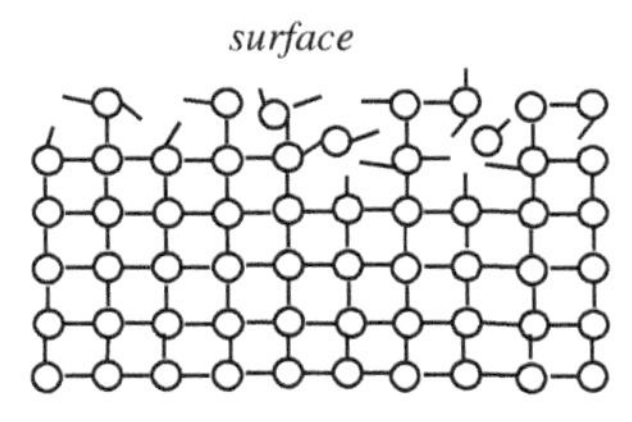


**Fig. 2.3.** Illustration of the disturbed interatomic bonds structure near the surface of the single-crystal semiconductor material.

According to the rules of quantum mechanics, the energy of electrons in a single atom is restricted to discrete levels. The discrete energy levels associated with single atoms are broadened into valence and conduction bands. Due to the different arrangements of atoms at the surface than in the bulk and the resulting different thermodynamic characteristics, positions of energy levels that can be occupied by an electron represented by the band structure are not the same at the surface and in the near-surface region as they are in the bulk of semiconductor samples. This difference manifests itself in the bending of the energy bands in the near-surface region (Fig. 2.4). Band bending is related to the presence of space charge in this region needed to accommodate the departure from equilibrium caused by the abrupt termination of the crystallographic structure at the surface and reflects differences in energy distribution at the surface and in the bulk of semiconductor materials.

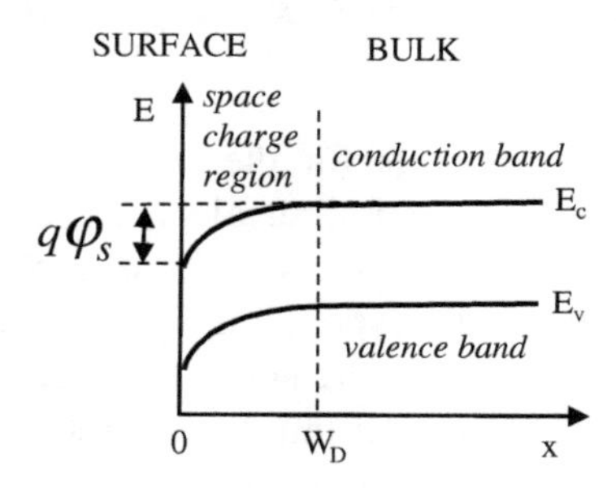


**Fig. 2.4.** Energy bands of a *p*-type semiconductor material showing a modification of the energy distribution at its surface expressed in terms of the surface potential $\varphi_s$.

The extent of the energy bands bending (downward bending in the case of *p*-type semiconductors) is expressed in terms of the surface potential $\varphi_s$ noted in Fig. 2.4. Surface potential, also referred to as surface barrier, varies depending on the condition of the surface as well as the composition of the ambient interacting with the surface (Zhang and Yates, 2012).

## 2.4 Crystal Structure of Semiconductor Materials in Surface Region

The process of surface formation impacts in a major fashion the crystallographic structure of semiconductor samples at the surface and in the near-surface region. As expected, the effect of surface-related phenomena is not the same in semiconductor materials featuring different crystallographic structures.

Figure 2.5 schematically illustrates two classes of solids identified based on their crystallographic structure. It is apparent from this illustration that the surface of highly ordered single-crystal materials (Fig. 2.5(a)) features less disordered broken bonds than in the case of polycrystalline materials (Fig. 2.5(b)). Distortion of the bond geometries in the near-surface region of the otherwise highly ordered single-crystal material is the main factor causing deterioration of its characteristics, including charge transport characteristics.

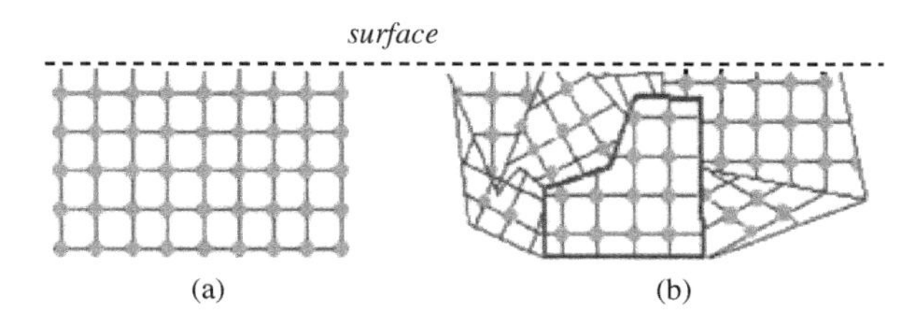


**Fig. 2.5.** Two-dimensional representations of crystallographic structure at the surface in the case of (a) single-crystal material and (b) polycrystalline material.

In the polycrystalline material (Fig. 2.5(b)), the charge transport characteristics are controlled predominantly by the boundaries between grains and the size of the grains. Consequently, the effect of the surface on the charge transport characteristics is relatively less pronounced in this case. Similar reasoning applies to amorphous materials where the charge transport across structurally disordered samples is dependent on the scattering phenomena across the volume, regardless of whether carriers are moving closer or further away from the surface.

The conclusion from the above comments is that in polycrystalline and amorphous materials, factors related to surface distortion have a limited effect on the overall characteristics of materials. In contrast, surface-related features have a pronounced effect on the overall performance of the material in device applications in the case of single crystals. Keeping this conclusion in mind, further discussion in this volume concerned with the characteristics and characterization of semiconductor surfaces is focused on single-crystal semiconductors.

**Crystal lattices.** The term "crystal lattice" is used here in reference to the repeated three-dimensional arrangement of atoms in the crystal.

Any single-crystal lattice comprises the elemental cells reproduced throughout the material. Elemental cells may appear in a variety of forms falling into seven basic classes. The key elemental semiconductor materials used to fabricate semiconductor devices, specifically silicon (Si) and germanium (Ge), belong to the cubic class of crystals in the variety which follows the crystal lattice of a diamond (Fig. 2.6(a)). In addition to the cubic class of crystals, some key compound semiconductor materials, most notably gallium nitride, GaN, crystallize in the hexagonal crystal configuration (Fig. 2.6(b)). Based on the illustrations in Fig. 2.6, one can envision differences in surface characteristics, including density and configuration of broken bonds at the surface, between semiconductor materials featuring different crystal lattices.

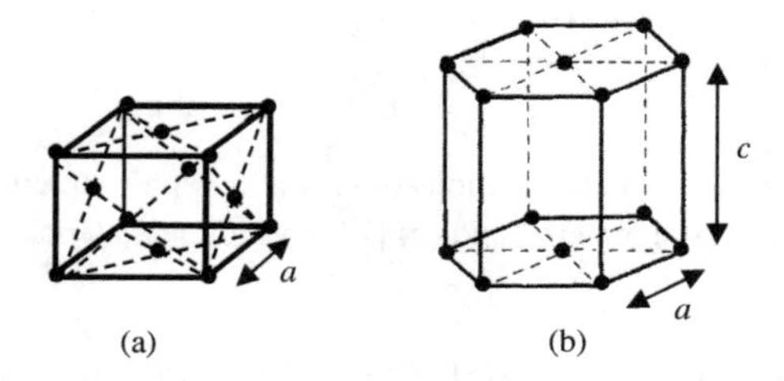


**Fig. 2.6.** Lattice constants in (a) face-centered cubic cell, *a*, and (b) hexagonal cell, *a* and *c*.

As can be expected, the cell-based structures illustrated in Fig. 2.6 are severely distorted in the near-surface region. Among other factors, the extent of distortion depends on the crystallographic plane defined by the Miller indices along which a single-crystal ingot is cut into the wafers as well as on the way the created surface is finished.

**Miller indices.** By connecting the points on the main crystallographic axes of the single-crystal material, various crystallographic planes within the crystal can be identified using Miller indices. In the case of cubic cells, Miller indices involve three digits, each either 1 or 0, for instance, (100) or (111), which define how the plane intersects the main crystallographic axes of the crystal (Fig. 2.7). In the case of hexagonal cells, the additional index needs to be used to describe all crystallographic planes.

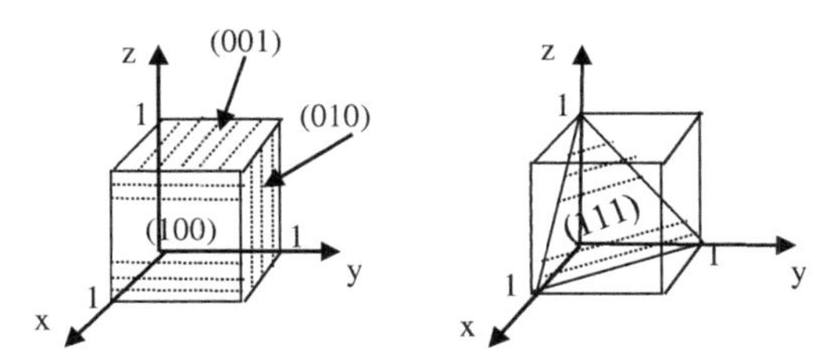


**Fig. 2.7.** Use of Miller indices to define various crystallographic planes in the cubic crystal.

As it can be concluded from the illustration in Fig. 2.7, the density of unsaturated bonds on the surface varies depending on the surface orientation defined by Miller indices. For instance, a cubic crystal with a surface along the (100) plane (Fig. 2.7(a)) will feature different densities of unsaturated bonds on the surface than in the case of a cubic crystal cut along the (111) plane (Fig. 2.7(b)) (Aspnes and Handler, 1966). Also, other parameters of crystals vary depending on surface orientation (Chongsawangvirod and Irene, 1991). Following the established procedures, Miller indices are being used to identify the crystallographic orientation of various surfaces throughout this book.

**Defects in single-crystal lattice.** Even the highest quality single-crystal semiconductor materials used in the manufacture of semiconductor devices contain structural and compositional imperfections referred to as defects. In this discussion of special interest are defects in the near-surface region introduced during device manufacturing procedures rather than those resulting from the crystal growth process. The former are predominantly point defects appearing in the crystal in the form of atoms missing from the lattice (vacancies, or clusters of vacancies forming voids), often expending into the near-surface region, alien atoms substituting for the host atoms or excess atoms at the surface located interstitially (interstitial defect), or substitutionally (substitutional defect), as shown in Fig. 2.8.

Other types of defects observed in single crystals, including line defects, planar defects, and volume defects are mostly the results of the single-crystal growth process and cannot be controlled using standard surface processing procedures. Engineering of the surfaces of semiconductor crystals towards minimization of the density of process-induced structural defects at the surface and in the near-surface region is a major objective in surface processing of single-crystal materials used in the manufacture of semiconductor devices.

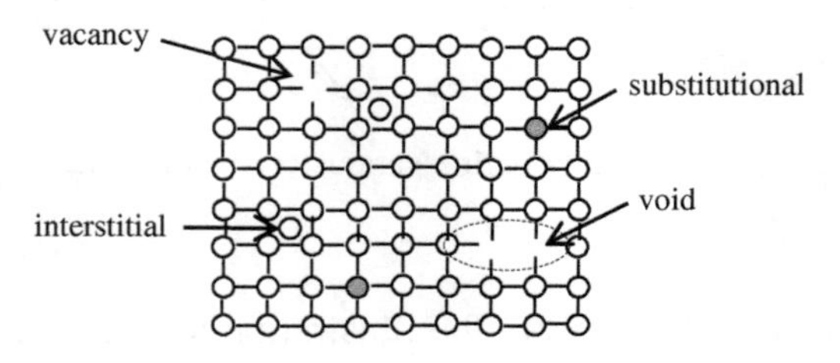


**Fig. 2.8.** Examples of point defects in the cubic lattice of a single-crystal semiconductor.

This observation applies in particular to device operation which depends on the characteristics of the semiconductor surface (see discussion in Chapter 5). This is because, in the case of a high concentration of defects in the single-crystal semiconductor, their adverse effect dominates over the intrinsic physical properties of the crystal. On the other hand, if the intrinsic physical properties of the crystal are inferior, as it is in the case of amorphous materials, intricate deficiencies related to point defects at the surface do not come into play. Additional comments regarding the effect of defects at the surface of the single-crystal materials on the operation of semiconductor devices can be found in Section 4.3 of Chapter 4.

## 2.5 Chemical Composition of Semiconductor Materials in Surface Region

As a reminder, Table 2.1 shows a part of the Periodic Table of Elements often referred to as a Semiconductor Table of Elements. It identifies elements in group IV featuring distinct semiconductor properties and elements in groups II, III, V, and VI used to synthesize compound semiconductors. Elements in group IV, namely carbon (C), silicon (Si), and germanium (Ge), are the only elements displaying semiconductor properties. The last one in this group, tin (Sn) features properties which in elemental form make it unsuitable for semiconductor device applications.

**Table 2.1.** Semiconductor Table of Elements.

| II | III | IV | V | VI |
|---|---|---|---|---|
| | B<br>boron | C<br>carbon | N<br>nitrogen | O<br>oxygen |
| | Al<br>aluminum | Si<br>silicon | P<br>phosphorus | S<br>sulfur |
| Zn<br>zinc | Ga<br>gallium | Ge<br>germanium | As<br>arsenic | Se<br>selenium |
| Cd<br>cadmium | In<br>indium | Sn<br>tin | Sb<br>antimony | Te<br>tellurium |

Among elemental semiconductors, silicon is the most widely used. What makes silicon a broadly used semiconductor material is a combination of abundance, relatively low cost of obtaining high-quality crystals as compared to other key semiconductors, high-quality native oxide $SiO_2$, and device-compatible electronic properties. Germanium, which was a semiconductor material of choice in early device applications including the very first demonstration of transistor action, features higher charge carrier mobilities, holes in particular, but is less chemically resistant and mechanically stable than silicon. Carbon is the third element in group IV of the Periodic Table (Table 2.1), which in 3D, single-crystal configuration crystallizes as diamond and in 1D configuration is known as graphene.

The difference in the way the surfaces of silicon and germanium respond to thermal oxidation exemplifies the effect of the chemical composition of a material on the characteristics of its surface. In the case of silicon, its reaction with oxygen at elevated temperatures forms strong Si–O bonds and leaves its surface covered with a layer of very stable silicon dioxide $SiO_2$. In contrast, the thermal oxidation of germanium leaves on its surface unstable, water-soluble germanium oxide, $GeO_2$.

Different effects influence the surface characteristics of binary semiconductor compounds comprising elements A and B (Fig. 2.9(a)) featuring different properties. As an example, the thermal oxidation of silicon carbide, SiC, produces gaseous carbon oxides (CO and $CO_2$) removing carbon atoms from the surface and leaving it covered with silicon dioxide. Effectively then, thermally oxidized SiC forms on its surface carbon-free silicon dioxide $SiO_2$ which when removed by etching, leaves the surface deficient in carbon (element B in Fig. 2.9(b)).

Considering gallium arsenide (GaAs) as another example, the difference is in the significantly higher vapor pressure of arsenic as compared to gallium. When exposed to temperature above 600°C, arsenic evaporates leaving behind gallium which collapses into the powder-like material. At the reduced pressure conditions, even at the lower temperature, evaporation of arsenic from the near-surface region can be observed leaving the surface of GaAs crystal arsenic deficient (element B in Fig. 2.9(b)). The situation is similar in the case of cadmium sulfide (CdS) where due to the different response to elevated temperature between elements forming a compound, excessive sulfur vacancies are created in the near-surface region (Nishimura, 1991).

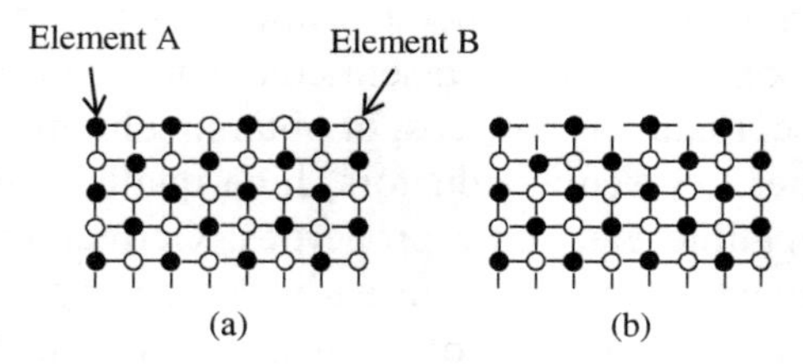


**Fig. 2.9.** Surface features of a binary semiconductor compound formed using elements A and B displaying different properties: (a) before the process and (b) after the process involving conditions to which elements A and B respond differently.

The above discussion emphasized inherent differences between surface chemistry responses to external conditions (temperature and pressure) in the case of elemental and binary semiconductors as well as other more complex ternary and quaternary semiconductor compounds. In this book, the discussion of the characteristics and the characterization of semiconductor surfaces is concerned predominantly with the elements from group IV of the Periodic Table (Table 2.1), with emphasis on single-crystal silicon.

## 2.6 Electrical Conductivity of Semiconductor Materials in Surface Region

Electrical conductivity $\sigma$ of semiconductor materials is determined by the nature and strength of the bonds between the atoms comprising the solid which defines the availability of free electrons responsible for its electrical conductivity. Only electrons free to move under the influence of the electric field $\mathcal{E}$ can contribute to electric current $J = \sigma \mathcal{E}$. In the case of *n*-type semiconductors, electrical conductivity $\sigma$ depends on the concentration of electrons $n$ and their mobility $\mu_n$ ($\sigma \sim n \mu_n$).

Similarly, the concentration of free holes and their mobility define the electrical conductivity of *p*-type semiconductors. Various aspects of the impact of the abrupt termination of the crystal structure at the surface on electron concentration and mobility, and thus on the electrical conductivity of semiconductor materials are discussed in this section.

First, however, while considering the surface effect on the electrical conductivity of semiconductors, the difference between current flowing in the direction normal and parallel to the surface needs to be noted. As the discussion in Chapter 5 will remind us, there are semiconductor devices operation of which is based on the control of the current flowing between ohmic contacts in the direction normal to the surface (Fig. 2.10(a)). In this case, while potentially affecting the properties of the ohmic contacts being an integral part of such devices, the surface condition has a minor effect on the flow of charge carriers comprising the current.

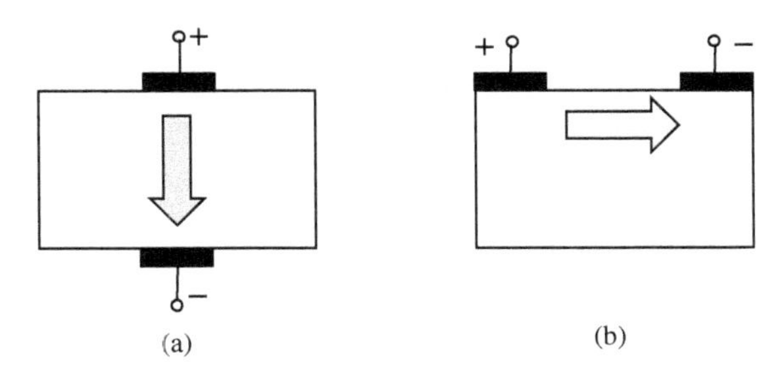


**Fig. 2.10.** Flow of current in a semiconductor device in the direction (a) normal to the surface and (b) parallel to the surface.

In contrast, in devices such as, for instance, metal–oxide–semiconductor field-effect transistor (MOSFET), the current flows between ohmic contacts in the direction parallel to the surface (Fig. 2.10(b)), in which case the surface effects play a pronounced role in defining the performance of the device. Considering the scope and objectives of this book, further discussion in this section will be concerned primarily with the latter.

The concentration of electrons, which is the main contributor to the conductivity of *n*-type semiconductors, can be permanently changed by orders of magnitude through the modification of their chemical composition by adding foreign elements acting as dopants. Doping can be implemented either during the crystal growth process or during device manufacturing by ion implantation or diffusion. In either case, the condition of the surface plays a negligible role in determining the outcome of the doping process. Instead, modifications of the electrical conductivity in the near-surface region of semiconductors are determined by the effects linking the concentration of free electrons and their mobility with the extent of the disruption of the crystallographic structure and potential contamination of the near-surface region.

**Generation–recombination processes.** While doping processes establish a foundation upon which the electrical conductivity of semiconductor materials is based, significant changes in the concentration of charge carriers are driven by the generation and recombination of free electrons and holes within their near-surface region.

As a reminder, generation is the process of free charge carrier formation in semiconductors resulting from the electron in the valence band acquiring enough energy through heating or illumination, for instance, to overcome the energy gap and transition to the conduction band leaving a free hole in the valence band (Fig. 2.11(a)). Once in the conduction band, an electron is free to move in the presence of an electric field and contribute to the current in semiconductor samples.

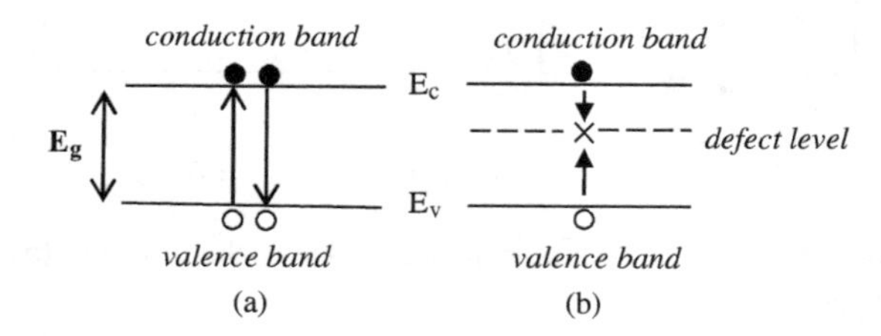


**Fig. 2.11.** (a) Band-to-band generation–recombination processes in single-crystal semiconductor materials are altered by (b) defects in the crystal lattice introducing defect levels in the bandgap.

The process of recombination, effectively leading to the annihilation of free charge carriers, is opposite to the generation of the same and results in the release of energy resulting from the electron transition from the higher energy level in the conduction band to the lower energy level in the valence band where they recombine with holes. The energy resulting from recombination can be released either in the form of light or in the form of the heat dissipated into the lattice of the semiconductor depending on whether the energy gap of the semiconductor is direct or indirect.

A point being made here is that generation–recombination processes are affected by the structural deficiencies and contamination of material, introducing additional energy levels in the semiconductor bandgap referred to as generation–recombination centers (Fig. 2.11(b)). In the case of high-quality crystals, such as those used in semiconductor device manufacturing, the density of generation–recombination centers at the surface is higher than in the bulk of the material. These centers play a role in defining the electrical conductivity in the near-surface region of semiconductor materials.

**Minority carrier lifetime.** The time between minority charge carrier generation and recombination is referred to as minority carrier lifetime, $\tau$, and is denoted as $\tau_n$ and $\tau_p$ for the electrons and holes, respectively. The minority carrier lifetime is strongly dependent on the density of defects acting as recombination centers (Kurjata-Pfitzner, 1980), and thus is a good measure of the quality of the crystal. From the point of view of the discussion in this book, it is essential that the methods used to monitor minority carrier lifetime allow separation of the effects controlling carrier lifetime in the bulk of the wafer and at its surface.

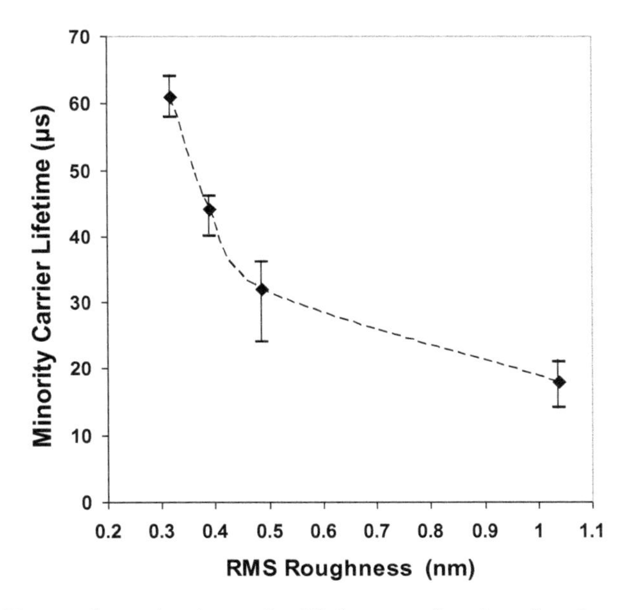


**Fig. 2.12.** Near-surface minority carrier lifetime as a function of surface roughness of germanium wafer.

As an illustration of these effects, Fig. 2.12 (Drummond *et al.*, 2009) shows changes in the minority carrier lifetime $\tau$ measured using the photoconductance decay (PCD) method discussed in Chapter 7 as a function of the increasing roughness of the measured Ge wafer. The results shown indicate a strong dependence of the minority carrier lifetime on surface roughness expressed in this figure in terms of the root mean square (RMS) roughness discussed further in Chapter 4, and as such, justify the use of the minority carrier lifetime as a measure of the condition of the surface and near-surface region of semiconductor materials.

**Electron mobility.** Free electrons carrying an electric charge and moving in semiconductor materials are subject to scattering resulting from the collisions and electrostatic interactions with the host and dopant atoms in the lattice, as well as with structural defects and contaminants. All these interactions result in the alterations of the charge carrier's movement in semiconductor materials. Quantitatively, these alterations are collectively reflected in the changes in the value of electron mobility, $\mu_n$ ($cm^2/Vs$), which in addition to the electron concentration $n$ controls the electrical conductivity $\sigma$ of semiconductor materials ($\sigma \sim n\mu_n$).

The feature of the charge carrier mobility of importance to the discussion in this *Guide* is that electron mobility in the structurally disturbed near-surface region of the semiconductor sample is strongly dependent on the condition of the semiconductor surface in the vicinity of which the current is flowing. As an example, Fig. 2.13 shows a decrease in carrier mobility with gradually increasing surface roughness similar to the minority carrier lifetime considered above (Drummond *et al.*, 2009).

Another experimental result testifying the same effect shows significantly reduced electron mobility in the channel of the MOSFET formed on the purposely roughened Si surface (Ohmi *et al.*, 1991). The interactions controlling the mobility of charge carriers in semiconductor materials in the vicinity of their surface, including surface roughness, are further considered later in this book.

**Electron transport.** In order to carry the electric current, charge carriers need to be set in motion. Factors affecting the transport of charge carriers in semiconductors are recalled here because the flow of electric current in semiconductor materials is strongly affected by the features of the surface region, particularly in the case of current flowing in the direction parallel to the surface in its vicinity (Fig. 2.10(b)). Under normal conditions in

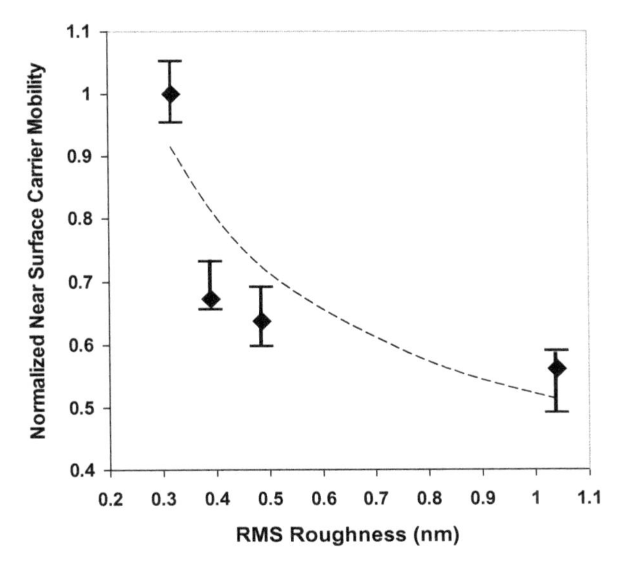


**Fig. 2.13.** Near-surface charge carrier mobility as a function of surface roughness of germanium wafer.

terms of geometrical confinement of semiconductor materials, which otherwise may bring quantum effects into play, there are two mechanisms of charge carrier transport that can generate electric current resulting from the net flow of electrons in the *n*-type semiconductor material.

First is a drift (drift current) which is a movement of electrons driven by the electric field and which is determined by the electrical conductivity of the material $\sigma$ and electric field $\mathcal{E}$ ($J = \sigma\mathcal{E}$). The strong dependence of $\sigma$ on the features associated with a near-surface region of semiconductor is a reason why the density of drift current $J$ under the same electric field $\mathcal{E}$ is expected to be different in the bulk of the material and at its surface.

The second mechanism, which does not require an electric field and which is specific to semiconductors, is based on charge carriers' diffusion (diffusion current). In this case, the flow of carriers is driven by concentration gradient, or in other words, non-uniform distribution of charge carriers in semiconductor materials. The current in this case flows in the direction of the lower concentration region until a uniform distribution of carriers is reached. In addition to the concentration gradient, its density also depends on the diffusion coefficient of charge carriers which in turn depends on their mobility. As determined above, the mobility of charge carriers is different in the near-surface region of the semiconductor and in its bulk.

What the above-mentioned comments indicate is that regardless of whether the current in semiconductor materials is driven by the electric field (drift current) or by the concentration gradient (diffusion current), their densities in the near-surface region of a single-crystal semiconductor sample and in its bulk are expected to be different.

**Velocity saturation.** The velocity of electrons moving in semiconductors under the influence of an electric field, referred to as drift velocity, increases with the increasing electric field and saturates at a certain maximum value. Saturation occurs because of the excessive scattering of charge carriers drifting in the semiconductor lattice with very high velocity.

Both saturation velocity and electric field at which velocity saturates are material parameters which are different in semiconductor materials featuring different spatial distributions of atoms in the crystal lattice. Considering the alteration of the spatial distribution of atoms in the crystal lattice at the surface, the saturation velocity and electric field at which the velocity saturates are lower in the near-surface region than in the bulk of semiconductors.

**Conclusions regarding electrical conductivity.** As the discussion in this section indicates, electrical conductivity at the surface of high-quality single-crystal semiconductor materials is lower than in their bulk. The extent and nature of the difference depend upon the factors discussed in this chapter, as well as on the nature of interactions with external influences to which surface is subjected and which are considered in the following chapter.

## 2.7 Effect of Surface Non-Planarity and Nanoscale Confinement

Starting wafers used in the manufacture of conventional semiconductor electronic and photonic devices feature atomically flat surfaces. In some cases, a departure from this rule is needed to ensure improved performance of such devices (Ruzyllo, 2007). In the case of solar cells or photodetectors, for instance, surface texturing is applied to minimize the reflection and maximize the transmission of the light across the surface (see discussion in Chapter 5). Other than that, surface geometry does not affect in a major way the operation of photonic devices which is typically controlled by the *p–n* junction located at a certain distance from the surface. The same observation applies to electronic devices in which current is flowing across the bulk of the wafer between contacts on its top and back surfaces (Fig. 2.10(a)).

In contrast, in electronic devices in which the current flows in the direction parallel to the surface in its vicinity (Fig. 2.10(b)), the surface is an active part of the device operation which is also influenced by the way the surface is shaped. An example here is a "fin"-based field-effect transistor, FinFET in short, upon which advanced digital integrated circuits are based.

In FinFET technology, which came about as an extension of planar MOSFET technology facing limits of transistor scaling, the transistor's channel is positioned vertically on its side assuming a fin-like shape (Fig. 2.14(a)). The result is an increased area of the transistor's gate, and thus increased density of the transistor's gate capacitance as compared with planar configuration, at the reduced area on the chip surface occupied by the transistor.

The FinFET geometry is typically accomplished by etching grooves in single-crystal silicon wafers and forming fin-shaped structures upon which transistors are being formed. The test structure processed on the silicon-on-insulator (SOI) wafer and representing this type of configuration is shown in Fig. 2.14(b).

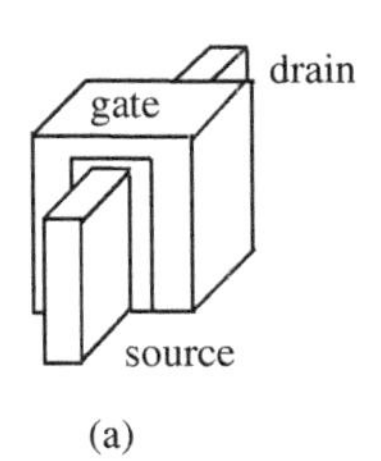


(a)

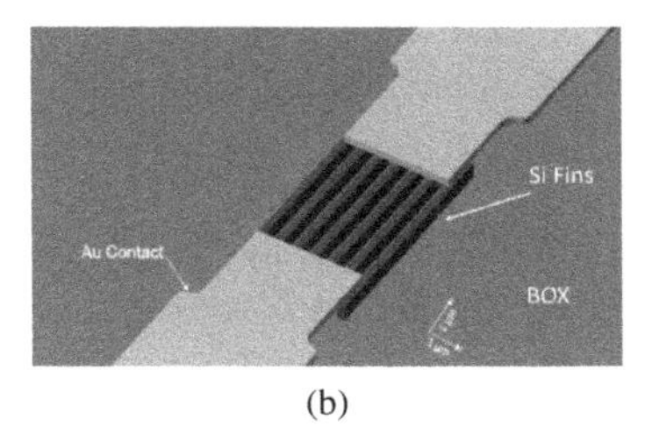


(b)

**Fig. 2.14.** (a) Schematic representation of the FinFET and (b) grooves etched in single-crystal silicon in silicon-on-insulator (SOI) wafer forming a structure upon which FinFETs are formed.

While improving the scalability of the MOSFET circuitry, introduction into the transistor structure of vertical surfaces and related post-etch damage brings about new challenges in terms of the control of the condition of the surface and near-surface region. The nature of these challenges manifests itself in the lower minority carrier lifetime in the fin-shaped surface region denoted as Finned SOI in Fig. 2.15 ($\tau = 15.6$ $\mu$s) as compared to planar SOI ($\tau = 45.2$ $\mu$s) shown in the same figure (Drummond *et al.*, 2017).

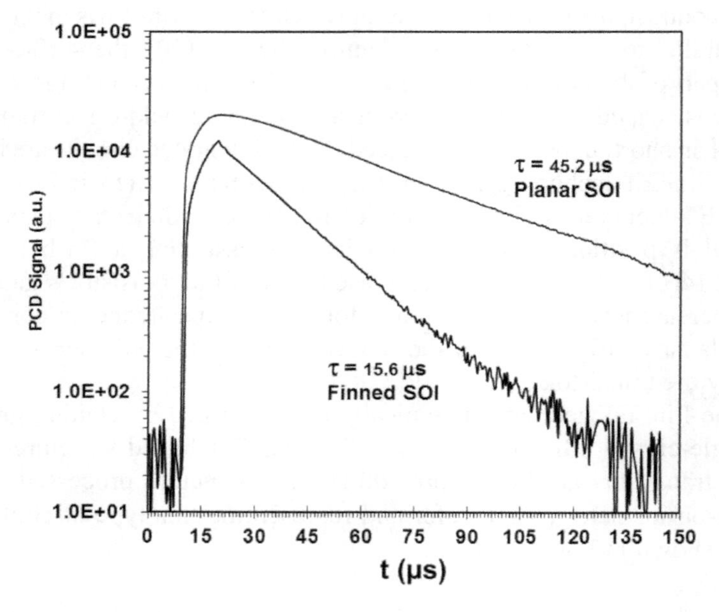


**Fig. 2.15.** Minority carrier lifetime reduction in the fin-shaped surface region (Finned SOI) as compared to planar SOI.

In some other MOS field-effect transistor (MOSFET) applications, the non-planarity of the gate structure is implemented for reasons other than transistor scaling. For instance, the vertical configuration of the gate region in the vertical sleet field-effect transistor (VESTFET) results in the superior to planar MOSFET on/off currents ratio (Maly *et al.*, 2011).

Another purposely introduced modification of semiconductor surface geometry is implemented in power MOSFETs. Here, the challenge is to ensure an increasingly large area of the transistor gate structure so that the transistor can handle more power without increasing the area of the wafer surface it occupies. To accomplish this goal, substrate wafer can be etched to form U-well-shaped features into which the transistor's gate is built forming what is known as UMOSFET (Fig. 2.16) (Suliman *et al.*, 2001).

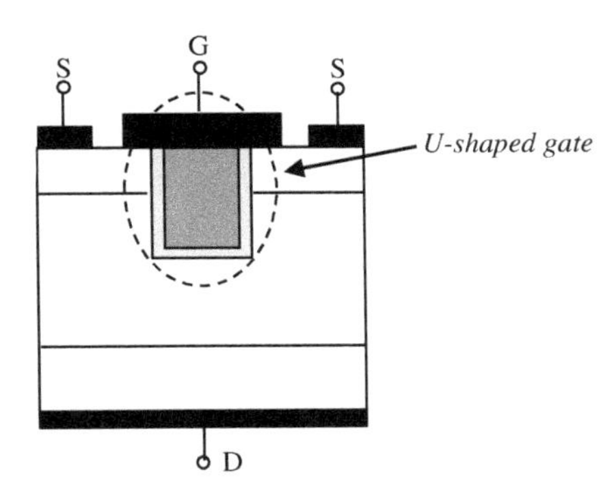


**Fig. 2.16.** MOSFETs featuring the U-shaped gate as an example of the surface geometry alterations for the purpose of improving the performance of power transistors.

In the cases of transistor configuration shown in Figs. 2.14 and 2.16, processing and characterization of non-planar silicon surfaces is calling for innovative solutions (Wu *et al.*, 2002). Some of them are considered later in this volume in Chapters 6 and 8.

The basic properties of semiconductor materials are independent of the volume of the sample until their geometry is reduced to the atomic scale, or in other words, to the nanoscale. The nanoscale confinement of semiconductor materials brings about changes in their properties which are no longer defined by the rules of classical physics but instead are subject to the rules of quantum physics. Examples of nanoconfined material systems include, for instance, 2D graphene, in which case interactions with gaseous ambient are different than in the case of conventional carbon surfaces (Gosh *et al.*, 2008), 1D nanowires, and 0D quantum nanodots used, for instance, as a downshifting medium when deposited on the surface of silicon solar cells (Sabeeh *et al.*, 2019).

Considering that the discussion in this book is focused on the characteristics and characterization of the surface and near-surface region separate from the characteristics of the bulk of the semiconductor material, the discussion of the surface interactions in the nanoscale material systems in which strictly surface-related effects cannot be isolated is beyond the scope of the discussion in this book and as such is not pursued. The same concerns micro-electro-mechanical system (MEMS) devices in which the mechanical characteristics of silicon determine their operations and the processing of the surface is limited to the cleaning operations and etch residue removal in the beam release processes (Erdamar *et al.*, 2008; Allen, 2005).

# Chapter 3

# Interactions of Semiconductor Surfaces

## 3.1 Introduction

As indicated earlier, the discussion in this *Guide* is focused on the issues related to the practical aspects of semiconductor surface engineering involved with semiconductor device manufacturing rather than those concerned with free-surface science. The purpose of this chapter is to review interactions the surface of the semiconductor wafer may be exposed to during device fabrication, wafer storage and handling, as well as in some other cases, for instance, in the case of solar cells, during device operation. The discussion of these topics is driven by the need to establish a knowledge base necessary to follow elaborations in the remaining chapters of this volume, in which processing and characterization of semiconductor surfaces are considered in the context of semiconductor device processing. In the instances where the discussion of the surface characteristics should be related to a specific semiconductor material, silicon is used as a material of reference.

The chapter begins with introductory comments concerned with the effect condition of the surface has in the case of processes such as epitaxial deposition. Then, physical and chemical characteristics of semiconductor surfaces are considered, followed by the sections addressing specifically interactions with ambient, visible light, electric and magnetic fields, as well as interactions with high-energy electrons and ions. Furthermore, the effect of the wafer's exposure to the elevated temperature on the surface and near-surface regions and its dependence on the way thermal energy is delivered to the wafer is considered.

## 3.2 Physical and Chemical Characteristics of Semiconductor Surfaces

As an introduction to the discussion in the remaining parts of this chapter, this section briefly considers what is being understood here under the terms physical (Kingston, 2016) and chemical characteristics of semiconductor surfaces. The finely finished surface of single-crystal silicon (Fig. 2.1) is used here as an example representing the nature of phenomena discussed with recognition of the inherent differences between the physical properties of crystalline and non-crystalline silicon surfaces.

Before continuing the discussion concerned with the physical and chemical characteristics of the surface, let's briefly consider the process of epitaxial deposition as a prime example of a strong correlation between the condition of the surface and the outcome of the process. In the case of the undisturbed, either physically or chemically, surface of the substrate, the growth of the epitaxial film reproducing a crystallographic structure of the substrate is possible (Fig. 3.1(a)) (Ruzyllo, 2020). However, when the surface of the substrate is disturbed either physically or chemically or both, the growth of the single-crystal film reproducing a crystallographic structure of the substrate is not possible (Fig. 3.1(b)).

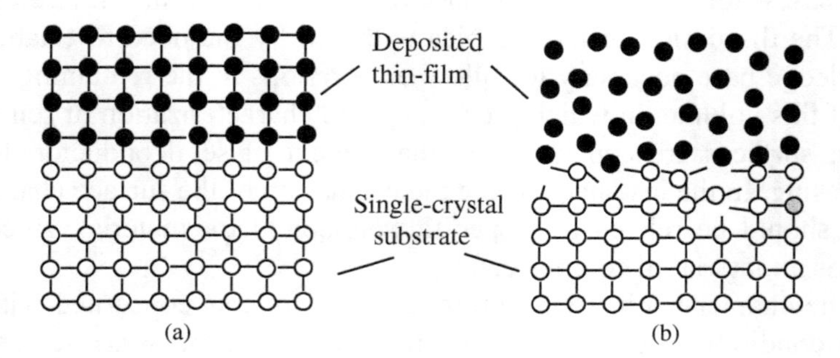


**Fig. 3.1.** Simplified illustration of the growth of (a) epitaxial layer on a defect-free surface and (b) structurally disordered film on a defective surface.

The distinction between the physical and chemical characteristics of the surface adopted in this discussion is somewhat arbitrary as the physical and chemical characteristics of the solid surfaces are intertwined and their effect on the electrical and optical properties of the surface cannot be always clearly separated. The exception is a case of semiconductor surface in a high-vacuum environment in which chemical interactions with an ambient do not come into play and physical features associated with the structurally disturbed surface and near-surface region are the manifestation of the departure of surface features from the ideal condition of the entirely undisturbed surface.

As a follow-up, Fig. 3.2 illustrates in a simplified fashion disturbances affecting the characteristics of silicon surfaces in the vacuum environment (no chemical interactions) which are concerned with structural features of the surface, including point defects (Fig. 2.8), altered bond geometry, and electric charges (Benedek and Toennis, 2018).

Resulting from the disturbance of single-crystal semiconductor material's physical structure at the surface shown in Fig. 3.2 are changes in the distribution of electric charge at the surface and in the near-surface region. Unless purposely passivated, the semiconductor surface remains electrically active, and the density of the resulting electric charge is a measure of its condition. The origin of surface charge and its measurements are further discussed in the later parts of this book.

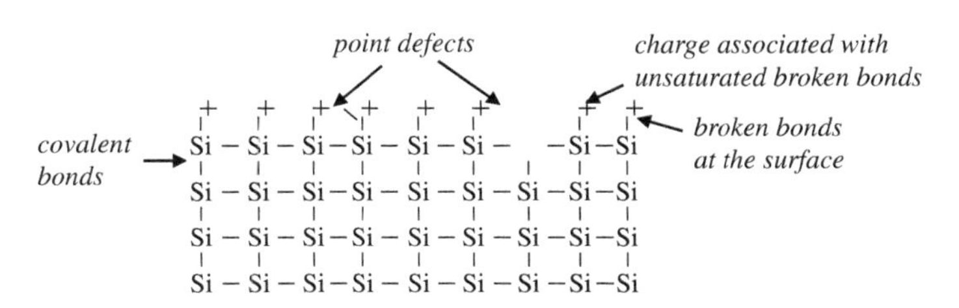


**Fig. 3.2.** Covalently bonded silicon with broken (dangling) bonds and displaced atoms at the surface.

The physical properties of the semiconductor's surface and near-surface region, including surface charge, affect the electron transport characteristics in the region near the surface. Surface defects are responsible for trapping and scattering of moving carriers accounting for the shortened minority carrier lifetime and reduced mobility of charge carriers in the vicinity of the surface. Operations aimed at the minimization of the effect of the deficient physical properties of the surface on the performance of the final devices include thermally driven gettering and epitaxial extension considered earlier (Fig. 1.4).

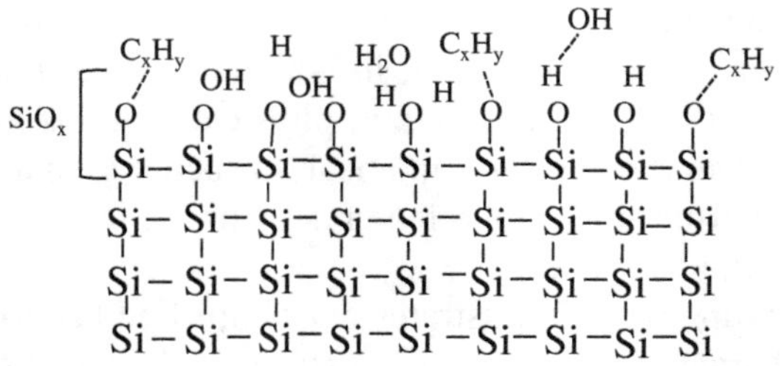

**Fig. 3.3.** Simplified illustration of chemical condition of silicon surface exposed to ambient air.

As indicated in the following section, in the device manufacturing or laboratory environment, physical alterations of the surface structure are accompanied by the changes in its chemical makeup resulting from interactions with an ambient. Whether it is a cleanroom ambient air or rinsing water, species containing oxygen, hydrogen, and hydrocarbons get physically attached to the surface, as shown in an illustrative fashion in Fig. 3.3.

In the case of silicon considered here, because of its very high affinity to oxygen, oxygen reaction with broken silicon bonds at the surface is inevitable. As a result, changes in the chemical composition of the surface proceed toward the formation of the spontaneously grown, ultra-thin, quasi-stoichiometric oxide of silicon, $SiO_x$, on the surface (Fig. 3.3).

In addition to oxygen interacting with the silicon surface, the dissociation of water molecules from the moisture in the ambient air and adsorbed at the surface into $H^+$ and $OH^-$ is a part of the reaction. Furthermore, organic components of ambient air in the form of hydrocarbons $C_xH_y$ are commonly found on silicon surfaces exposed to such air. In the case of other semiconductors, organic compounds with contribution from moisture are mainly responsible for the alteration of the chemical composition of the surface exposed to ambient air.

In a situation where uncontrolled variations of the chemical composition of the surface need to be prevented, the surface can be subjected to the passivation process which renders it chemically inactive. In the case of silicon surface, passivation is commonly accomplished by enforcing, using dedicated chemical treatments, the termination of unsaturated silicon bonds at the surface with hydrogen (Fig. 3.4). Silicon forms with hydrogen bond, which by the virtue of its strength prevents the reaction of silicon with other gaseous species commonly present in the ambient air including oxygen.

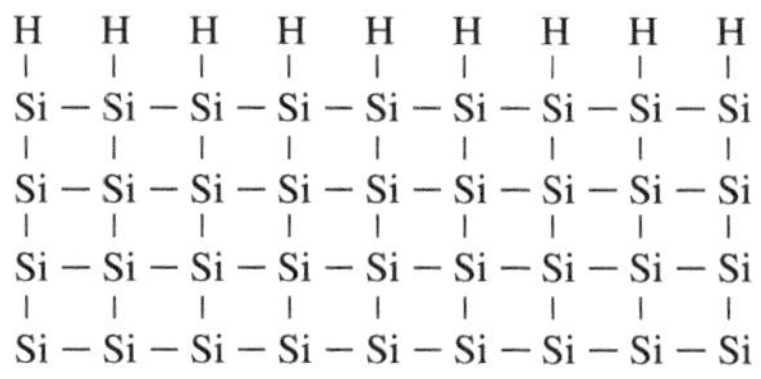

```
H    H    H    H    H    H    H    H    H
|    |    |    |    |    |    |    |    |
Si – Si – Si – Si – Si – Si – Si – Si – Si
|    |    |    |    |    |    |    |    |
Si – Si – Si – Si – Si – Si – Si – Si – Si
|    |    |    |    |    |    |    |    |
Si – Si – Si – Si – Si – Si – Si – Si – Si
|    |    |    |    |    |    |    |    |
Si – Si – Si – Si – Si – Si – Si – Si – Si
```

**Fig. 3.4.** Hydrogen-passivated silicon surface.

Also, the thermal oxidation of silicon resulting in the formation of a thin layer of silicon dioxide, $SiO_2$, on its surface prevents uncontrolled chemical reactions involving the surface. On the other hand, however, the surface of the silicon dioxide itself is prone to interactions with an ambient, destabilizing its condition.

The key point in this discussion is that whatever semiconductor material is of interest, and however it is being used in device applications, control over physical and chemical characteristics of its surface is of essence. The discussion in the following section and later in this chapter considers possible controlled and uncontrolled interactions involving semiconductor surfaces during device fabrication.

## 3.3 Interactions with Ambient

Except for the semiconductor wafers enclosed in the high-vacuum environment, the surfaces of such wafers are always subjected to interactions with the surrounding ambient. They may involve interactions during wafer processing with ambient featuring composition designated to perform specific tasks such as surface cleaning or etching, as well as interactions with ambient air during wafer storage and wafer handling (Torek, Mieckowski, *et al.*, 1995).

Unless surfaces are passivated (for instance, hydrogen-passivated silicon surface shown in Fig. 3.3), any exposure to ambient, whether gaseous or liquid, causes changes in the chemical makeup of the surface of semiconductor material. Furthermore, any exposure of semiconductor surfaces to the ambient, which is accompanied by the delivery of energy to the surface, causes changes in its physical characteristics. Thermal energy associated with elevated temperature treatments or interactions with high kinetic energy ions bombarding the surface are the example. Alterations of surface characteristics, either chemical or physical in nature, change the fundamental features of the surface such as surface energy and surface charge, and are readily detectable and monitored using methods selected among those discussed in Chapter 7.

In this section, alterations of the condition of the silicon surface resulting from the interactions with either gaseous or liquid ambient are considered. The discussion is focused on the impact and control of surface interactions with the ambient during semiconductor device fabrication and is not concerned with the fundamental science behind physical and chemical phenomena that occur at the interface between two phases, including solid–gas and solid–liquid interfaces.

Interactions of solid surfaces with gases are at the core of manufacturing processes converting single-crystal semiconductor materials into functional devices. Besides, even when not subjected to device-building interactions, semiconductor surfaces are also exposed to gaseous ambient during wafer shipping, storage, and handling. Therefore, an understanding of the nature of these interactions is needed to better control the surface characteristics in semiconductor device engineering.

Operations performed during device manufacturing on semiconductor wafers, and with its top surface directly exposed to the environment, in various ways alter the physical characteristics of the surface or enforce changes in its chemical composition, or both. For the sake of this discussion, let's consider the formation on the silicon surface of the thin film of its native oxide, silicon dioxide $SiO_2$, by means of chemical vapor deposition (CVD) and thermal oxidation.

The CVD processes are based on the chemical reaction in the gas-phase product of which is a solid deposited on the exposed surface (Fig. 3.5(a)). In the case considered here, the deposition process involves a gas-phase reaction between oxygen and silane ($SiH_4$) carried out at the elevated temperature needed to stimulate the desired reaction. The process forms a thin layer of $SiO_2$ on the silicon surface, but because gaseous reactants do not chemically react with silicon atoms on the surface, it does not alter in a major way the chemical composition and physical structure of the surface and the near-surface region.

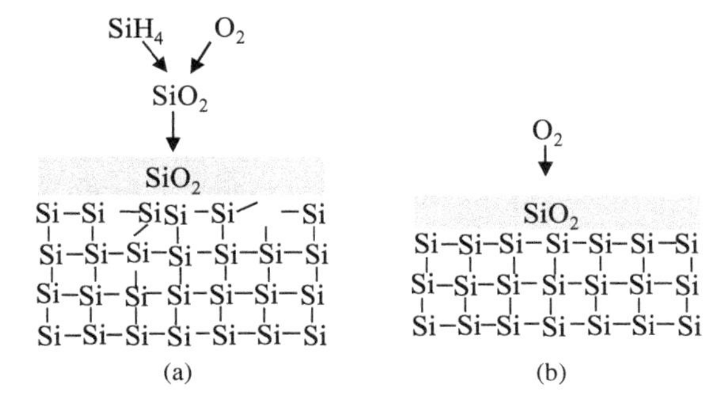


**Fig. 3.5.** Examples of surface–gas interactions involved in the formation of a thin layer of $SiO_2$ on silicon surface by (a) deposition using chemical vapor deposition (CVD) and (b) thermal oxidation of silicon.

In contrast, during the growth of the thin layer of $SiO_2$ on the silicon surface by means of high-temperature thermal oxidation in which case the silicon surface is exposed to gaseous oxygen only. The reaction of thermal oxidation of silicon results in the $SiO_2$ growing into the surface region and converting its most damaged parts into thermal silicon dioxide $SiO_2$ (Fig. 3.5(b)).

Another way gaseous species may react with the surface involves subtractive operations where as a product of chemical reactions between gaseous species and surface atoms, volatile compounds are formed. Resulting etching of material in the near-surface region leaves the surface altered both physically and chemically. The ways in which etching processes affect the surface of the etched single-crystal semiconductor material are further considered in Chapter 6.

To better understand the impact of the environment on semiconductor surfaces, the concepts of adsorption and desorption of chemical and physical species on the solid surfaces are briefly considered.

**Adsorption.** Adsorption is the process by which molecules of gas or liquid adhere to the surface of a solid, referred to as adsorbent, and form on its surface a thin film of the adsorbate (Fig. 3.6). Adsorption is a complex surface phenomenon, dependent upon several factors including porosity, roughness, topography, and electrical conductivity of the surface of adsorbent, as well as chemical composition and electrostatic properties of adsorbates in relation to adsorbent. Also, the temperature of the substrate and the pressure of the ambient gas play a role in defining a balance between adsorption and desorption processes. Overall, both adsorption and desorption control interactions of semiconductor surfaces with the ambient involved in semiconductor device manufacturing and as such are further considered below.

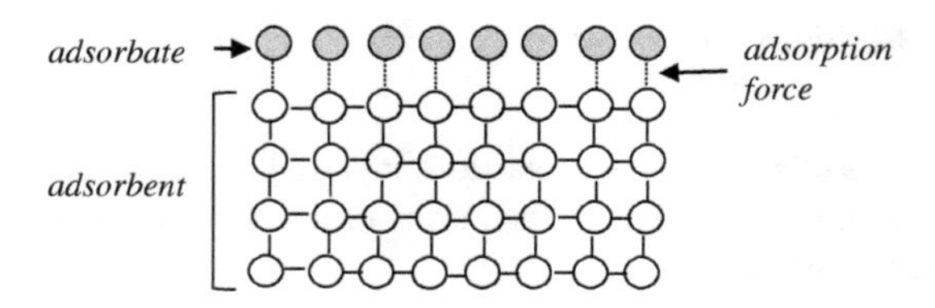


**Fig. 3.6.** Illustration of adsorbate, adsorbent, as well as adsorption forces, either physical or chemical.

The process of adsorption needs to be seen in terms of physisorption and chemisorption depending on the nature of phenomena controlling interactions between adsorbate and adsorbent. The former takes place when adsorbate molecules are attracted to the surface but are not involved in any chemical interactions with surface atoms. This type of adsorption is typically controlled by the weak van der Waals forces (Lüth, 2010) which are the most common in nature forces upon which attraction of the gas molecules to the solid surfaces occurs. Since the van der Waals' forces are universal, the surface of an adsorbent does not show any preference for an adsorbate which means that the physisorption process is independent of the chemical makeup of the adsorbent surface. What it also means is that the activation energy for physisorption is relatively low, and physisorption is a reversible process. Moreover, the process of physisorption is exothermic in nature and as a result, physical adsorption occurs more readily at the lower temperature of the substrate. On the other hand, reduced pressure weakens adsorption forces and promotes desorption of the gas molecules.

In contrast to physisorption, chemisorption involves the formation of chemical bonds with the surface, which results in the alteration of its electronic structure. Also, in contrast to physisorption, chemisorption is highly specific and occurs only if there is a local environment allowing the formation of chemical bonding between adsorbent and adsorbate such as, for instance, the Si–O bond during the thermal oxidation of silicon.

As it may be intuitively expected, the activation energy for chemisorption is higher than in the case of physisorption, and chemisorption occurs slowly. In some cases, the physisorption of a gas adsorbed at low temperatures may change into chemisorption when temperature increases. Also, during prolonged storage in ambient air, originally physiosorbed gaseous species may form chemical bonds to the surface because of the processes catalyzed by the moisture in the ambient air.

Due to these features, the process of chemisorption is practically irreversible without dedicated operations which are breaking chemical bonds linking adsorbate and absorbent and causing desorption of the former.

**Desorption.** As the name indicates, desorption is the reverse of adsorption and is the effect through which physiosorbed or chemisorbed molecules of gaseous substances are released from the surface.

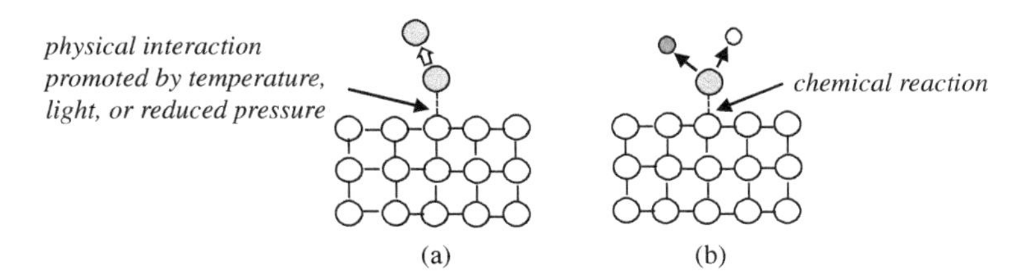


**Fig. 3.7.** Desorption from the surface of (a) physiosorbed and (b) chemisorbed molecules.

In the case of physiosorbed species, the process of desorption is typically driven by thermal stimulation, combined or not, with the reduced pressure or illumination needed to weaken the physical force attracting gaseous molecules to the surface (Fig. 3.7(a)). In the case of chemisorbed molecules, appropriately formulated compounds delivered to the surface chemically react with adsorbate molecules and convert them into volatile or water-soluble species (Fig. 3.7(b)).

Overall, interactions illustrated in Fig. 3.7 play an important role in the surface processing operations in semiconductor device manufacturing and as such will be further considered in the follow-up discussion in this book.

In this context, of interest are interactions taking place when the semiconductor surface, silicon in the case exemplified here, is exposed to the cleanroom air or storage ambient at room temperature and atmospheric pressure. The result is a gradual change in the chemical composition of the silicon surface promoted by the adsorption of moisture, oxygen, and organic compounds originating from the ambient air. As shown in Fig. 3.8, this change is reflected in the changes of contact angle (also referred to as wetting angle, the concept of which is considered in the following chapter) on the silicon wafers featuring initially either hydrophobic or hydrophilic surface characteristics during prolonged wafer storage in a polypropylene container (Tsai *et al.*, 2003).

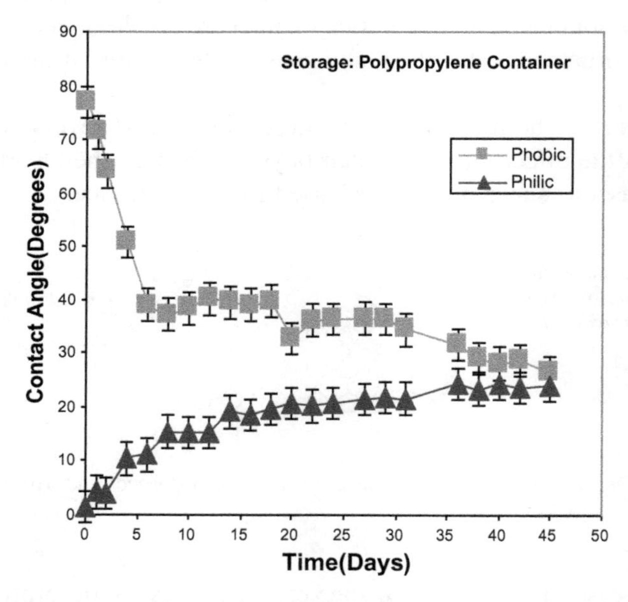


**Fig. 3.8.** Changes in the chemical makeup of initially hydrophobic and hydrophilic Si surfaces expressed by contact (wetting) angle variations as a function of time of wafer storage in a polypropylene container.

In the former case, originally close to 80°, the wetting angle gradually decreases in the course of a few weeks long exposure to ambient air inside the storage container to about 25°. Simultaneously, reflecting changes in the chemical condition of the surface, the contact angle on the originally hydrophilic surface increases from about 0° and reaches over time the same value as in the case of the originally hydrophobic surface (Fig. 3.8).

During the surface interactions illustrated in Fig. 3.8, sometimes referred to as surface "aging," changes in the chemical makeup of silicon surface that are taking place need to be controlled, and if needed, prevented or reversed. Other than oxygen and moisture, changes in surface condition are the result of surface interactions with organic compounds unavoidably present in ambient air during wafer handling in the cleanroom as well as during wafer storage and shipping. Reversal of these changes comes down to the removal of organic contaminants from the surface, which can be accomplished using the methods discussed in Chapter 6, including techniques specifically developed for this purpose such as rapid optical surface treatment, or ROST (Kamieniecki, 2001; Danel *et al.*, 2003; Shanmugasundaram *et al.*, 2004).

An example of the effect of ambient air interactions with semiconductor surfaces on the characteristics of semiconductor devices involves appropriate processing of the surface in the course of ohmic contact formation. As discussed in Section 1.6, by allowing the undisturbed flow of electric current from and into the device, ohmic contacts are needed to establish electrical communication with external circuitry. Contacts of this type must feature series resistance $R_s$ as close to zero as possible which, as alluded to earlier, requires adequate processing of the semiconductor surface prior to the metal deposition step.

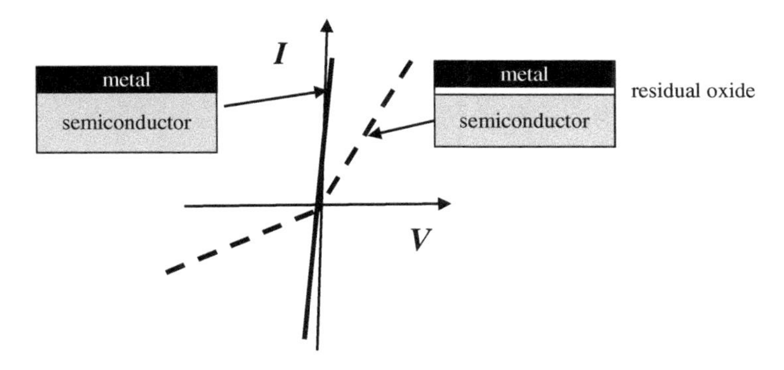


**Fig. 3.9.** Current–voltage (*I–V*) characteristics of the contact featuring direct physical contact between metal and semiconductor and contact with ultra-thin interfacial oxide layer resulting from interactions with an ambient.

The goal of surface treatments at this point is to ensure that the surface of semiconductor and metal or metal alloy forming ohmic contact remains after deposition in the intimate physical contact without any interlayer. As shown in Fig. 3.9, only in such cases, the current–voltage characteristic of the contact features high slope and lack of dependence on the applied voltage, indicating the characteristics of the ohmic contact.

The presence of an interfacial film, even in the nanometer thickness range, formed as a result of the exposure of the semiconductor surface to ambient air before metal deposition, makes contact to lose its ohmic characteristics and introduce series resistance, adversely affecting the performance of the device, of which metal–semiconductor contact is a part (Fig. 3.9). For this reason, the processing step removing any film spontaneously formed on the surface as a result of interaction with an ambient prior to contact deposition is an integral part of the ohmic contact fabrication.

Another factor that needs to be considered in the context of the discussion of interactions of the semiconductor surface with an ambient is related to the effect of the ambient and its composition on the electrical conductivity of the surface itself. Semiconductor surfaces between ohmic contacts feature finite electrical conductivity which, if not controlled, can adversely affect device performance by allowing leakage current to flow between contacts on the surface (Fig. 3.10).

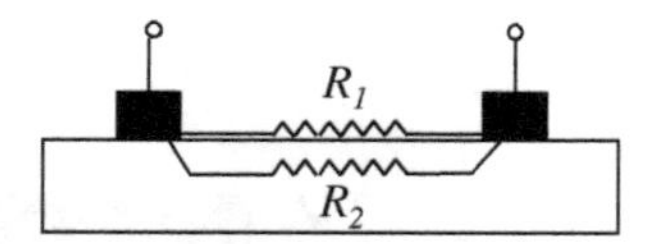


**Fig. 3.10.** Parasitic surface conduction path responsible for surface leakage featuring resistivity $R_1$ between ohmic contacts on semiconductor surface.

Unless purposely rendered electrically inactive by forming a layer of insulator, or through the processes of surface passivation, the surface provides a conduction path featuring resistance $R_1$ between two contacts remaining at the different potential, as indicated in Fig. 3.10.

Under certain conditions, resistance $R_1$ can be lower than the resistance $R_2$ of the contacts and the semiconductor's near-surface region between contacts. Depending on the condition of the surface, the conduction path allowing electron hopping between surface defect sites or ionic conduction in the case of moisture adsorbed at the surface can account for the uncontrolled leakage current between contacts on the surface.

Surface leakage current resulting from the interactions considered above is a parasitic effect which needs to be kept under control to ensure proper operation of the semiconductor device. For this reason, no part of the surface in the structure of a discrete semiconductor device or an integrated circuit can remain unprotected and electrically active. Most commonly, leakage due to surface conductivity is prevented by forming an adequately thick (Ruzyllo, Jakubowski, *et al.*, 1978) layer of an insulator between contacts and assuring surface resistivity $R_1 >> R_2$ (Fig. 3.10).

To prevent uncontrolled variations in the chemical composition of semiconductor surfaces in ambient air at atmospheric pressure, wafers between processing steps in the device fabrication sequence are stored in the inert gas ambient or in airtight containers at below atmospheric pressure. Alternatively, semiconductor surfaces can be subjected to the processes of surface passivation discussed earlier. Treatments of this nature render semiconductor surfaces chemically inactive and assure the stability and reproducibility of their characteristics.

In conclusion of the discussion in this section, constructive aspects of the sensitivity of semiconductor surfaces to the ambients altering their chemical characteristics need to be underscored. First of all, if needed, the chemistry of the surface can be altered in a controlled fashion to accomplish specific goals such as surface functionalization discussed later in this volume. Furthermore, the sensitivity of the semiconductor surface to the changes in the composition of the ambient is being exploited in the construction of certain kinds of devices performing sensing functions.

## 3.4 Interactions with Light

In addition to the impact on electronic properties, physical and chemical features of the surface also play a role in determining how it interacts with light. Physical characteristics such as surface structural defects or surface roughness affect reflectivity of the surface as well as absorption of the energy of light near the surface which impacts charge carrier generation and recombination processes in this region.

The effect of chemical properties of the surface on the material interactions with light is also pronounced. However, its consideration in the simplified, introductory fashion assumed in this discussion using silicon surface as an example would not serve a useful purpose. This is because a broad range of semiconductor materials used to manufacture photonic devices and the diverse purposes these devices serve do not allow drawing generally valid conclusions regarding the correlation between the chemical composition of the surface and the performance of any given semiconductor photonic device.

In the case of semiconductor photonic devices, interactions with light need to be considered from two different viewpoints. The first is concerned with the situation involving semiconductor photodiodes, photodetectors, and solar cells, whose role is to convert the energy of light into electrical signals. In this case, the light generated by an outside source, for instance a sun, illuminates the semiconductor material and promotes the desired interactions within its structure.

Different considerations apply to light-generating devices such as light-emitting diodes (LEDs). Here, the generated light featuring wavelength corresponding to the energy gap of the semiconductor is emitted from the inside of the device and crosses the surface inside-out with as limited losses as possible.

In the case of light-converting devices, the energy carried by the light needs to get across the surface and penetrate the given semiconductor material up to the depth defined by the absorption coefficient determined for any specific wavelengths. In the case of silicon, for instance, visible light in the portion of the spectrum from 0.4 $\mu$m to 0.7 $\mu$m features absorption depth in the approximate range from 0.1 $\mu$m to 6 $\mu$m (Fig. 3.11). The light absorbed in the semiconductor and featuring energy higher than its energy gap $E_g$, which in the case of silicon ($E_g$ = 1.1 eV) means wavelength shorter than about 1.1 $\mu$m, generates photocurrent upon which the operation of solar cells and photodetectors is based. In this wavelength range, light penetrates silicon to a depth of tens of micrometers below the surface.

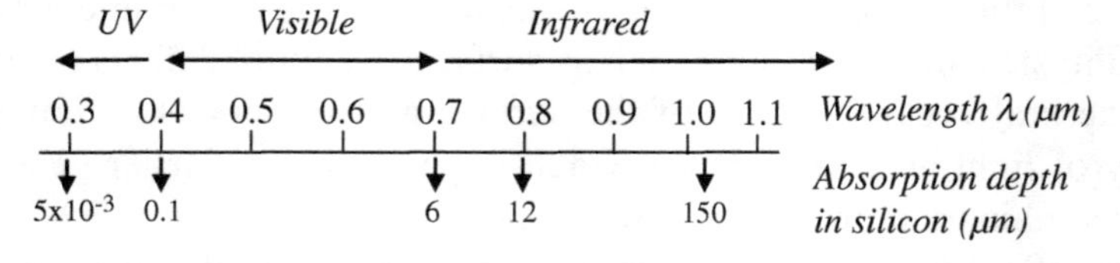


**Fig. 3.11.** Approximate values of absorption depth in silicon as a function of wavelength of light illuminating its surface.

Depending on the angle of incidence, refractive indices of the media involved (air and illuminated semiconductor material), as well as the condition of the illuminated surface, the beam of light impinging on the surface is subjected to interactions schematically illustrated in Fig. 3.12. The beam can cross the surface and the energy it carries will be absorbed in the close vicinity of the surface, which is a goal in the case of devices such as solar cells, or can be refracted with some losses, or can be totally reflected. The effect of refraction changes the direction of light passing through the boundary between two media featuring different optical properties expressed in terms of refractive indices $n$. Reflection in turn is the effect according to which light reaching the surface of an illuminated object returns to the same medium from which it was projected onto the surface.

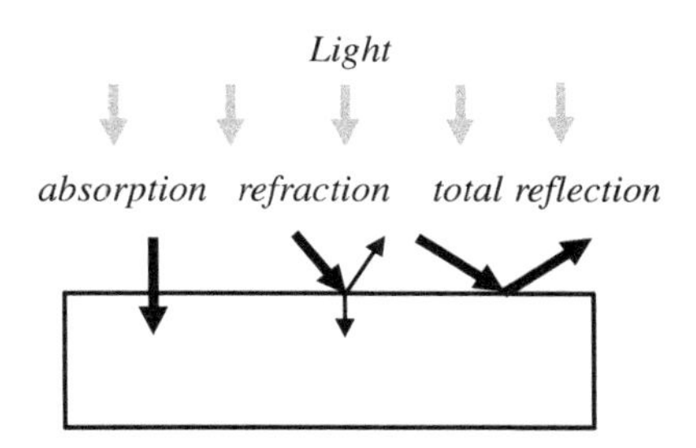


**Fig. 3.12.** Light from outside of the device illuminating its surface may be subjected to absorption, refraction, or total reflection.

From the point of view of the discussion in this volume, of importance is the dependence of reflection of light on the physical and chemical characteristics of the illuminated surface of semiconductor material, keeping in mind that the goal is to maximize the absorption of energy carried by the light. This is where the condition of the illuminated surface in devices such as solar cells, light detectors, or imaging devices comes into play. The goal is to reduce the reflection of light either by processing the outermost surface accordingly (see the discussion in Chapter 5), or by covering it with anti-reflection coating (ARC), or both.

Different considerations regarding interactions of semiconductor surfaces with light apply to the devices where light is generated within the semiconductor structure as it is in the case of light-emitting diodes (LEDs). Once generated, light needs to be projected to the outside of the device with minimal losses at which point, similar to the case of light from an outside source penetrating the semiconductor device (Fig. 3.12), interactions with the surface come into play. They are particularly important in the case of surface-emitting LEDs (SLED) and to a lesser extent in the case of edge-emitting LEDs (ELED), where light is emitted along the *p–n* junction rather than in the direction normal to the surface (Ruzyllo, 2020).

In the case of surface-emitting LEDs, the goal is to minimize losses of the intensity of light leaving the device by minimizing the adverse effect of internal reflection during the light passing through the surface inside out. It involves specialized surface treatments and changes of the surface shape not related to the outer surface processing and characterization, on which considerations in this book are focused.

## 3.5 Interactions with Magnetic and Electric Field

As indicated earlier, in any interactions of semiconductor material with energy coming from an outside source, the characteristics of the surface of the material exposed to such interactions play a role. In the following discussion, interactions of the semiconductor material with magnetic and electric fields are briefly considered.

Under normal conditions, the effect of magnetic and electric fields to which the semiconductor sample is exposed is not limited to interactions within the semiconductor's near-surface region and extends into the volume of the sample. While the electric field sets electrons in motion, the magnetic field alters their trajectories. The impact of both on the electron's flow dynamics in the surface region of the sample and in its bulk are different and the analysis of the nature of interactions involved is not a straightforward matter. For instance, the Hall Effect observed in semiconductor samples when the magnetic field is applied, which is commonly exploited in the measurements of electron mobility, does not differentiate between electron mobility in the bulk and in the near-surface region without additional experiments.

The effect of electric field in turn manifests itself in a variety of ways, depending on how and in what configuration the field is applied. The discussion in this *Guide* is focused on the effect of the electric field applied to the surface of semiconductor material for the purpose of controlling its near-surface electrical conductivity.

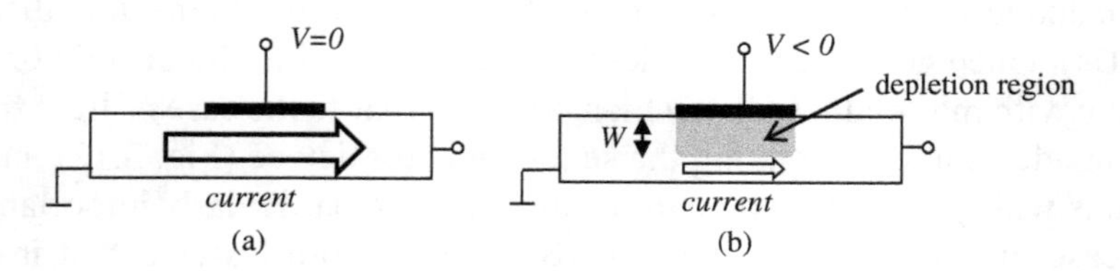


**Fig. 3.13.** Schematic illustration of the field effect according to which electron drift current in *n*-type semiconductor sample (a) is dependent upon the negative potential applied to the surface of the sample and controlling the thickness of the depletion layer $W$ (b).

As shown in Fig. 3.13(a), with an electrode in the immediate vicinity of the surface of *n*-type semiconductor and no voltage applied ($V = 0$), the flow of the current in the near-surface region of *n*-type semiconductor remains unaffected or is affected very weakly. Application of the negative bias ($V < 0$ in Fig. 3.13(b)) causes repulsion of electrons in the direction normal to the surface, creates a depletion layer $W$ near the surface of the semiconductor, and alters the flow of charge carriers. The effect of the electric field on the electrical conductivity of semiconductors in the sample configuration shown in Fig. 3.13(b) is limited to the near-surface region with limited impact on current flow in the bulk of semiconductors.

The field effect illustrated in Fig. 3.13 is exploited in the operation of field-effect transistors with emphasis on metal–oxide–semiconductor field effect transistor (MOSFET) operation, which, as discussed in Chapter 5, is strongly affected by the condition of the surface of semiconductor substrate used.

## 3.6 Interactions with High-Energy Electrons and Ions

Semiconductor surfaces which are not covered with any other material are exposed to energy-carrying electrons mainly during surface characterization processes which are discussed in Chapter 7. Due to the extremely small mass and size, high-energy electrons penetrating the surface and promoting the desired interactions in the near-surface region have a limited effect on surface characteristics. Other uses of electron beams involving applications in semiconductor device fabrication include electron-beam (e-beam) evaporation where there are no interactions with electrons in the beam and semiconductor surface and electron beam lithography where e-beam writing a pattern in the layer of electron-sensitive resist does not interact with the surface of semiconductor wafer.

In contrast, ions featuring much higher mass than electrons, when arriving at the exposed surface carrying significant kinetic energy, may cause sputtering of the top layers of the material regardless of its crystallographic structure and composition. In some applications, ion sputtering is purposely implemented during device fabrication or in material characterization to accomplish specific material removal-related goals. In others, such as, for instance, reactive ion etching (RIE), ions are causing damage to the surface which then needs to be treated to minimize its potentially harmful effect on the performance of semiconductor devices. A more detailed discussion of the RIE effect on the surfaces exposed to this process can be found in Section 6.4 of this book.

A noted special case is the process of ion implantation during which relatively light dopant ions cause alterations of the near-surface crystal structure, but do not sputter off atoms from the top layers of the implanted wafer. Dedicated post-implantation annealing processes are applied to minimize the structural damage of the surface and near-surface region caused by ion implantation.

## 3.7 Interactions with Thermal Energy

Thermally stimulated processes are an integral part of semiconductor device technology. The impact of thermal energy on the characteristics of the surface and near-surface region of semiconductor material depends on its thermal conductivity and the way thermal energy is delivered to the sample. It also depends on whether it originates from an outside source or is generated within a semiconductor device, either discreet or integrated, as a result of its operation.

In terms of temperature, of interest is the device operation range denoted in Fig. 3.14. At the temperatures below T1, where dopant ionization controls electrical conductivity $\sigma$ of semiconductor material, and above T2, where band-to-band generation of electron–hole pairs comes into play, surface-related effects have no impact on the electrical conductivity of semiconductor material.

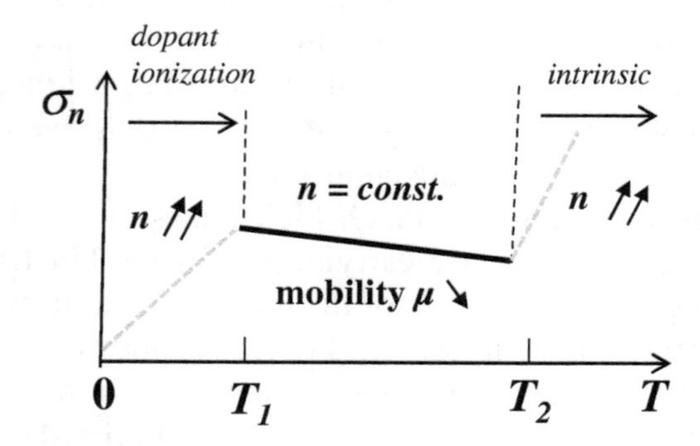


**Fig. 3.14.** Schematic qualitative representation of the changes of electron concentration $n$ and conductivity $\sigma$ of $n$-type semiconductor with temperature.

In the device operation range of *n*-type semiconductors, the concentration of electrons $n$ remains unchanged because dopant atoms are all ionized, and band-to-band carrier generation does not take place. The decrease of electrical conductivity shown in Fig. 3.14 comes from the decrease of electron mobility $\mu_n$ resulting from the increase of scattering of moving electrons on the atoms in the lattice, both host and dopant, all featuring thermally enhanced vibrations. As discussed earlier, electron scattering leading to reduced mobility is more pronounced at the surface than in the bulk of material because of the higher defect density in the former case. Therefore, when the semiconductor is in thermal equilibrium with an ambient and the temperature remains within the T1–T2 range, the near-surface-related phenomena affect the response of the device to the increased temperature, particularly in the case of devices operating based on the field-effect-induced changes of the charge distribution in the near-surface region.

The effect of thermal energy on the characteristics of the semiconductor surface depends not only on the temperature range but also on how thermal energy is delivered to the wafer during thermal treatments applied during device fabrication. The common option involves furnaces using conventional resistance heating elements during which the near-surface of semiconductor material and its bulk remain in thermal equilibrium (Fig. 3.15(a)).

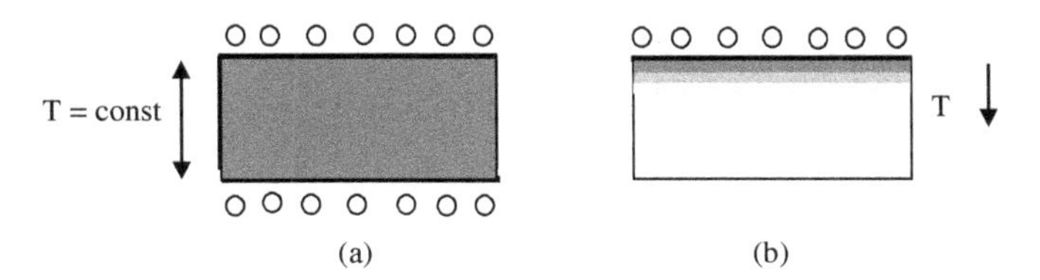


**Fig. 3.15.** (a) Distribution of temperature in the wafer during prolonged heating in the furnace and (b) in the wafer subjected to rapid thermal processing with heating elements on the front side of the wafer.

Different situations occur in the case of rapid thermal processing (RTP) using high-power lamps as heating elements on the front side of the wafer (Fig. 3.15(b)). In this case, during brief bursts of thermal energy, the exposed semiconductor wafer is not in thermal equilibrium. In such cases, its surface and near-surface region are more likely than the bulk of the same wafer to undergo thermally stimulated changes in its physical and chemical characteristics, particularly in the case of low thermal conductivity semiconductor materials.

A somewhat different case in terms of thermal energy interactions with semiconductor surfaces is represented by rapid optical surface treatments (ROST) referred to earlier in this chapter as a technique used to restore surface condition after prolonged exposure to ambient air. The ROST process which uses IR lamps as heating elements involves very brief (for instance 20 seconds) heating cycles raising the temperature of the wafer surface up to 200–300°C. Applications of this technique in semiconductor surface processing are further considered in Chapter 6.

Yet another variable affecting interactions of the semiconductor's near-surface region with thermal energy is concerned with excessive heat being generated within the working semiconductor device which may cause permanent damage to its surface and near-surface features. Most commonly, it happens because of high-density currents flowing across metal lines interconnecting elements within an integrated circuit. As in the case discussed above, here too the outcome depends on the thermal conductivity of materials used, which defines the ability of the device structure to dissipate generated heat. Also, designated cooling arrangements employed to control the chip's temperature come into play in this case.

To conclude the discussion in this chapter regarding interactions involving semiconductor surfaces, it needs to be noted that essentially any alterations of the features of semiconductor surfaces considered in this section, whether physical or chemical in nature, are reflected in one way or another in the changes of the electronic properties of the surface and near-surface region. Therefore, monitoring of the said electronic characteristics can be used to effectively detect changes in the characteristics of semiconductor surfaces. More in-depth coverage of this and related issues can be found in Chapters 7 and 8.

# Chapter 4

# Characteristics of Semiconductor Surface Defining its Condition

## 4.1 Introduction

The purpose of this chapter is to summarize the previously discussed characteristics of semiconductor surfaces and consider areas that can be used to assess their quality needed to assure adequate manufacturing yield and achieve expected performance of semiconductor devices. In other words, the discussion in this chapter is concerned with those characteristics of crystalline semiconductor surfaces which should be identifiable, well understood, and controllable to assure optimum operation of the practical semiconductor devices.

The discussion in this chapter is meant to reinforce a knowledge base needed to follow elaborations in the remaining chapters of this volume in which processing and characterization of semiconductor surfaces in the context of semiconductor device fabrication are considered. As before, in the instances where the discussion of the surface characteristics needs to be related to a specific semiconductor material, silicon is used as a material of reference.

A point that needs to be made in preparation of discussion in this chapter is that the characteristics of semiconductor surfaces considered are interrelated. As an example, surface energy and its wettability are altered by the presence of electric charges on the surface.

## 4.2 Surface Energy

Surface energy is a characteristic of the solid surface which quantifies the disruption of interatomic bonds that occur when a surface is created and affects its fundamental properties. As a key surface characteristic, it is a driving force behind initiation of interactions affecting solid crystalline surfaces such as surface reconstruction. In the case of semiconductor surfaces, of particular importance are their characteristics which directly affect semiconductor device processing, including the way the surface interacts with liquids and other materials with which it comes into contact. Discussion of the broadly understood surface energy is used here as a platform upon which other fundamental concepts describing the properties of the semiconductor surfaces of importance to the processing and operation of semiconductor devices are considered.

Surface energy is a physical characteristic which reflects the extent of disruption of interatomic bonds that occur when a surface is created. To explain the concept of surface energy, let's assume that semiconductor crystal is energetically in equilibrium, which means that its bulk bonding structure is undisturbed (Fig. 4.1(a)).

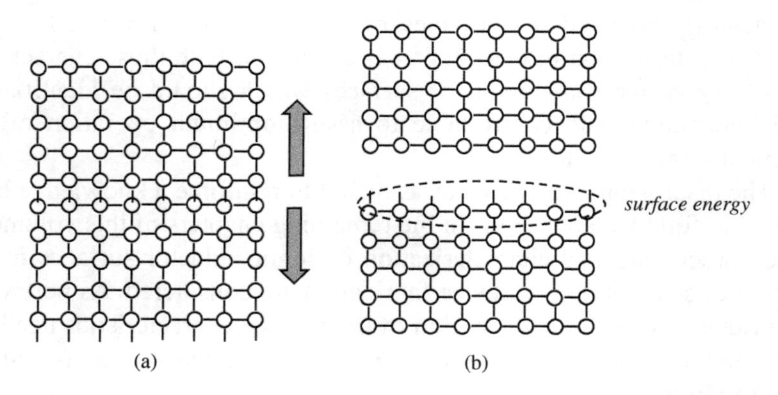


**Fig. 4.1.** (a) Process of cleaving single-crystal semiconductor sample creates the surface (b) featuring excess energy associated with the disruption of the bond structure and referred to as surface energy.

When the surface is created, for instance by crystal cleaving (Fig. 4.1(b)), the bond structure is disrupted and the energy is consumed in the process. This means that, on balance, more energy is now associated with the surface than with the bulk of the crystal. In other words, surface energy can be seen as the excess energy associated with the disruption of the bond structure at the surface. It quantifies the disruption of interatomic bonds that occur when a surface is created.

**Surface reconstruction.** Driven by the need to lower surface energy, perturbed crystallographic structure at the free surface spontaneously undergoes the process of reconstruction. Surface reconstruction relaxes atoms at the surface by modifying their position and bond configuration. As a result, the crystallographic structure of the atoms at the surface of the crystal is different from that in its bulk. In the semiconductor device manufacturing environment other than vacuum, the surface can be reconstructed into a lower energy state by adsorption, passivation, or oxidation. It means that the process of surface reconstruction affects surface energy during interactions of the surface with liquid and gaseous ambients during device fabrication.

In the case of the freshly formed surface under high-vacuum conditions, the unit surface energy of the sample formed (Fig. 4.1(b)) would be half of the material's energy of cohesion. In practice, however, the surface readily changes its form away from the "cleaved bond" model and its energy is often reduced by such processes as passivation (Fig. 3.4) or adsorption (Fig. 3.6).

The reason for our interest in the concept of surface energy is that by quantifying surface energy, we are gaining ability to control how other phases (gases, liquids), as well as other solids, interact with the surface of interest. For instance, energy of the surface defines mechanisms of nucleation of another material on it. This means that the ability to control the energy of the surface of semiconductor substrates gives us control over some key processes converting semiconductor materials into semiconductor devices. An efficient way to control surface energy is through alteration of the chemical composition of a crystalline semiconductor surface, which, between others, also changes the wettability of such a surface.

**Wetting (contact) angle.** One of the features of surface energy is that it is a measure of the tendency of the surface to repel a liquid. For instance, changes in surface energy affect the way the surface is interacting with water. In the case of low surface energy, featuring weaker attractive forces, the surface is hydrophobic, which means it tends to repel water causing its bidding on the surface, as indicated in Fig. 4.2(a) by wetting angle $\alpha$ (also referred to as contact angle) higher than some 60° and in some cases exceeding 90°. In contrast, hydrophilic surfaces which feature strong molecular interactions, and thus high surface energy, lend themselves to wetting as shown in Fig. 4.2(b) and are expressed by the wetting (contact) angle below 10° and approaching 0°.

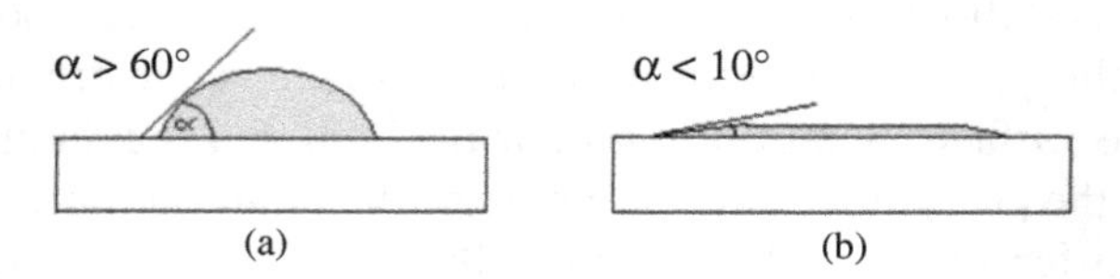


**Fig. 4.2.** (a) High wetting (contact) angle indicates hydrophobic surface, while (b) low wetting angle is an indication of the hydrophilic surface.

It needs to be stressed that the chemical composition of the liquid in contact with the surface also comes into play in determining the extent to which surfaces can be wetted. For instance, the surface repelling water may be readily wetted by the alcoholic solvent. Furthermore, there are chemical compounds, known as surfactants, formulated specifically for the purpose of altering surface energy in the desired fashion.

Lack of control over the changes in surface energy of semiconductor substrates may negatively affect the outcome of device manufacturing processes. Wetting angle is a very sensitive measure of surface characteristics affecting surface energy (Kwok and Neuman, 1999). For this reason, measurements of the wetting angle, or contact angle, are routinely used to monitor the condition of the semiconductor surface during device manufacturing. Examples of such measurements are considered later in this volume.

**Surface functionalization.** The term "functionalization" is used here in reference to the process which locally alters surface energy and, hence, its wettability (Tao and Bernasek, 2012). By locally altering surface wettability, the way it interacts with an ambient can be modified. The goal is to promote and control selective interactions between the surface and components of the gaseous or liquid environment to which the surface is exposed. The procedure can be implemented as a part of the bottom-up process involving, for instance, a self-assembly sequence.

As an example of the surface functionalization, the result of the procedure of locally converting originally hydrophilic surface into hydrophobic surface by means of UV exposure is shown in Fig. 4.3. This goal can be accomplished, for instance, by local exposure of the substrate surface to UV light which can be partially blocked off using a mechanical mask (Chao *et al.*, 2013).

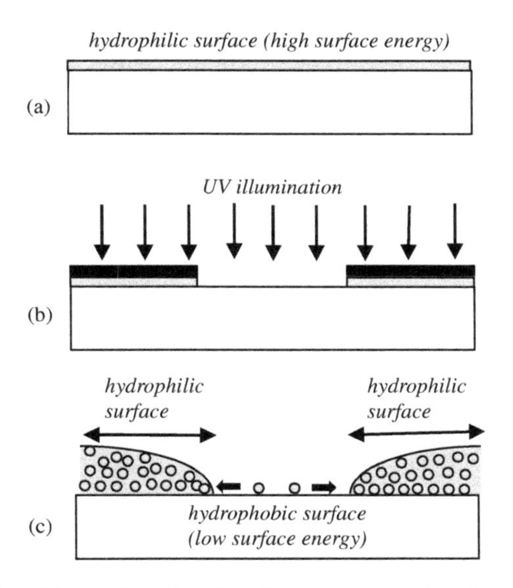


**Fig. 4.3.** Example of the surface functionalization process involving (a) conversion of the surface into uniformly hydrophilic, (b) exposure of the surface to UV light locally lowering surface energy followed by (c) deposition of the desired material in the vapor phase and causing its agglomeration in the hydrophilic areas.

As shown in Fig. 4.3, molecules contained in the vapor ambient coming into contact with functionalized surface are repelled from the low surface energy (hydrophobic) part, agglomerating on the higher energy (hydrophilic) part of the surface. What it means from the device fabrication perspective is that the geometry of the thin-film material is defined before its deposition by local alteration of chemical composition of the surface of the substrate rather than after deposition of photoresist followed by photolithography then etching and cleaning. In other words, by using the surface functionalization procedure, definition of the geometry of materials incompatible with a conventional photolithographic sequence and subsequent etching and cleaning is possible.

**Surface orientation.** As it can be intuitively expected, surface energy of the perfectly pure, free from any alien elements and maintained under ultra-high vacuum conditions, surface of the single-crystal semiconductor depends on its surface orientation. It means that under such conditions, surface energy is different for the differently oriented surfaces such as (100) and (111) surfaces of the same material (see Fig. 2.7). In practical situations, however, the surface unavoidably interacts with an ambient composition of which determines its chemical makeup. As a result, using adequately formulated chemical treatments, the surface energy of material can be varied according to the needs independently of its crystallographic orientation. This option can potentially be exploited in device manufacturing processes.

The essence of the above observation is that while surface orientation plays an important role in defining key properties of single-crystal material in semiconductor device applications, it has limited impact on surface energy during device manufacturing.

## 4.3 Surface Roughness

In semiconductor device engineering terminology, roughness refers to the disruption of planarity of semiconductor surfaces, which reflects distortion of a two-dimensional surface texture. As such, roughness is an inherently surface-related concept which is among the important material properties from the point of surface features having potentially adverse effects on the outcome of the semiconductor device manufacturing process. Following on earlier considerations in Chapters 2 and 3, in the remaining discussion in this *Guide*, of particular interest is surface roughness inflicted during device fabrication procedures such as surface cleaning or etching.

The roughness is represented by the difference between the highest and the deepest surface features. It can be as low as 0.05 nm in the case of the highest quality surface finish or as high as in micrometers range in the case of low quality or damaged semiconductor surfaces.

In semiconductor engineering terminology, surface roughness is commonly quantified by the vertical deviations of a real surface from its ideal form and is expressed as an arithmetic average of the absolute values $R_a$ of the microscopic peaks and valleys measured on the surface. Also commonly used to define surface roughness is calculated root mean square (*RMS*) of the measured microscopic peaks and valleys appearing on the surface.

Increasing values of RMS and $R_a$ are an indication of the increasing roughness of the surface. To exemplify the effect of the processing ambient on surface roughness, Fig. 4.4 shows an increase of roughness of germanium surface expressed in terms of RMS as a result of hours-long immersion in 100°C DI water (Drummond, Bhatia *et al.*, 2011).

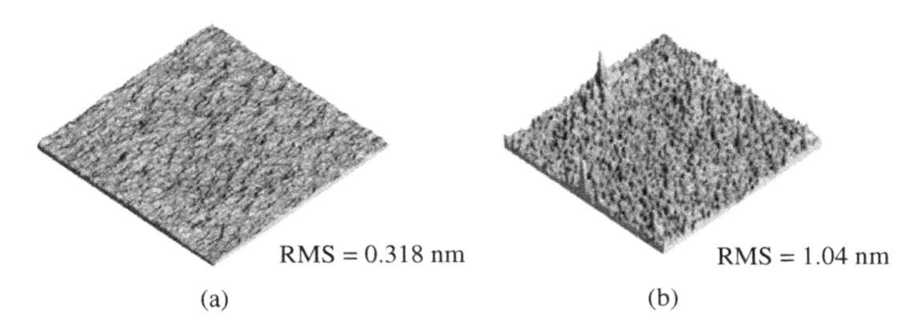


**Fig. 4.4.** Visualization of the surface roughness of germanium wafer (a) before and (b) after 6 hours immersion in boiling water.

While it may not be the case in other applications of conductive solids, in the manufacture of functional semiconductor devices, roughness of the surface of semiconductor substrates has in general a negative effect on the yield of the fabrication process as well as final device performance. The needs in terms of the surface physical smoothness, or surface roughness as defined earlier, vary depending on the type of device and materials used. In some cases, the surface may be textured on purpose, for instance, to reduce surface reflectivity in the case of solar cells. In some other situations, surface roughness may not be of significant concern as other deficiencies of the material or materials used may overshadow the adverse effect of the rough surface. In most of the other cases, however, surface roughness is a challenge of major concern in semiconductor processing. An adverse effect of surface roughness on the minority carrier lifetime and electron mobility in the near-surface region of semiconductor wafer was demonstrated earlier in Chapter 2 (Figs. 2.12 and 2.13).

From the perspective of discussion in this book, the issue of surface roughness in semiconductor device technology needs to be seen not only in terms of the roughness of the surface of the starting material but also in terms of the surface roughness inflicted by operations performed on the wafer during device manufacturing. For this reason, monitoring of surface roughness at various stages of the device fabrication process is an integral part of semiconductor device technology. This topic is addressed later in this volume.

## 4.4 Surface Defects

Among the structural defects of single-crystal materials, there are imperfections of the crystallographic structure of the single-crystal semiconductors specifically at the surface and in the near-surface region, referred to as surface defects. For the sake of clarity, surface defects are discussed here in terms of the substrate-originating defects and process-induced defects.

The first category is concerned with structural defects present on the surface of the single-crystal semiconductor wafer before it is subjected to device manufacturing processes. Those types of defects are related to the inherent characteristics of any given semiconductor and the constraints regarding its surface processing procedures. In the case of silicon, due to the mature surface processing technology (see Fig. 2.1), the surface defects are sufficiently well controlled to assure their limited negative impact on the device fabrication operations. This includes processes of denuded zone formation and epitaxial extension considered in Chapter 1 and illustrated in Fig. 1.4. Furthermore, in the case of silicon wafers featuring defective surface, the process of thermal oxidation during which oxygen from the ambient reacts at high temperature with silicon surface atoms converting into silicon dioxide the outermost defective parts of the silicon surface (see Fig. 3.5 in Chapter 3). The process known as sacrificial oxidation is then followed by the etching of the oxide, leaving silicon surface with significantly lower density of surface defects.

The challenges posed by substrate-originating defects are of a different nature than in silicon in the case of semiconductor materials featuring mechanical hardness preventing high-quality surface finish, such as in the case of silicon carbide substrate wafers. As shown by an electron micrograph of the SiC wafer surface in Fig. 4.5(a), severe mechanical damage which cannot be removed by the surface finishing operations is a common feature of SiC surfaces (Ridley *et al.*, 2000). If the geometry of surface scratch shown is pronounced in terms of its width and depth, then such a surface defect is reproduced in the epitaxial SiC film grown on such substrate, as illustrated in Fig. 4.5(b).

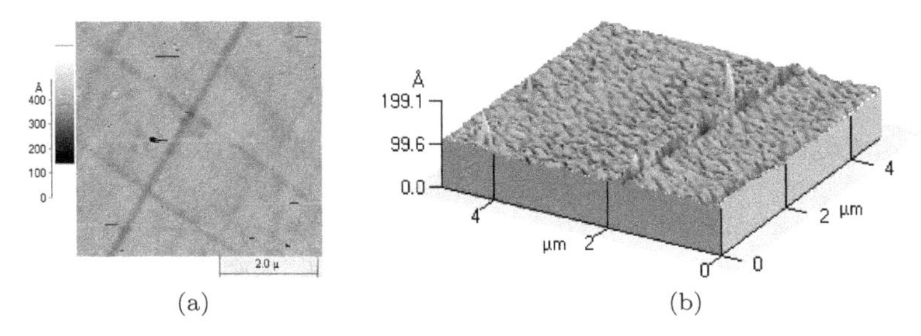


**Fig. 4.5.** (a) Mechanical damage of SiC surface reproduced as defect in (b) epitaxial film grown on such a surface.

The second category of surface defects concerns imperfections introduced to the surface and near-surface region of semiconductor substrates during device manufacturing procedures. Reduction of their density, detection, and characterization are among topics of interest for further discussion in this book.

Process-induced surface defects most commonly result from the bare surface exposure to the processes of etching and cleaning. Furthermore, they may result from surface exposure to electrically charged, mass-carrying species accelerated toward the surface such as ions during ion implantation.

Some of the process-induced defects, including those resulting from ion implantation, are well understood, predictable, and controlled with dedicated post-implantation annealing processes to which implanted wafers are routinely subjected. Other processes in which material removal from the surface is accomplished using ions carrying kinetic energy need to be determined and controlled for each set of conditions. For instance, the effect of reactive ion etching (RIE) requires dedicated treatments to neutralize surface damage (see discussion in Chapter 6).

Yet another example of etching-induced semiconductor surface damage during device manufacturing involves anisotropy of etching operations promoted by surface defects. This is because material removal by etching proceeds within the surface defect area at a rate different than in the defect-free portion of the surface exposed to etching chemistries causing permanent alterations of the surface topography. Moreover, potential impact of the surface defects on the homogeneity of dopant redistribution during doping processes performed on the wafer also needs to be noted.

Processes potentially damaging semiconductor surfaces are not limited to gas-phase interactions. Also, operations carried out in liquid-phase and involving aggressive chemistries constituting surface cleaning and etching chemistries may erode the surface, increasing its roughness.

Because of all the above-mentioned factors, no effort is being spared to minimize the density of surface defects in the materials used to fabricate semiconductor devices. To make it possible, a broad range of surface characterization techniques geared specifically toward detection as well as monitoring of surface defects is available. Related aspects of semiconductor surface engineering are considered in Chapters 7 and 8 of this volume.

## 4.5 Surface Potential

When a semiconductor surface is created, free electrons from the unsaturated broken bonds at the surface in Fig. 3.1 affect charge equilibrium in the material. Electrically active centers associated with broken interatomic bonds, or surface states, result from the disruption of periodicity of the lattice at the surface. Surface states interact with the sub-surface region of the sample by trapping free electrons or holes from the adjacent space charge region (Aspnes and Handler, 1966).

As a result of these effects, electrostatic characteristics of the surface differ from the equilibrium condition in the bulk of the sample. As indicated earlier in Fig. 2.4, the measure of semiconductor surface departure from the state of electrical neutrality in the bulk is a surface potential $\varphi_s$. By promoting repulsion or attraction of the free charge carriers, surface states enforce their separation which results in the appearance of an electric field $\mathcal{E} > 0$ in the near-surface depletion region featuring depth $W_D$ (Fig. 4.6).

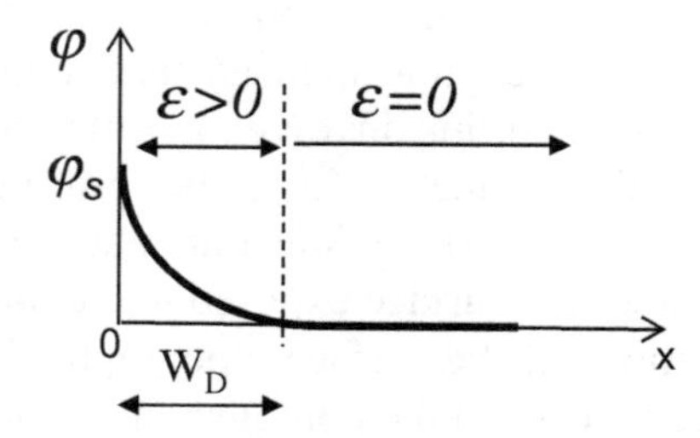


**Fig. 4.6.** Surface potential $\varphi_s$ represents the effect of the surface on the distribution of electric charges in the near-surface region, electric field in the surface region of semiconductor, and the depth of the depletion region $W_D$.

The surface potential $\varphi_s$ is a measure of the near-surface departure from electrostatic equilibrium in the bulk of semiconductor, where electric field $\mathcal{E} = 0$. An additional effect on the value of surface potential, acting toward its decrease or increase, comes from the potential applied to the surface of semiconductor, as shown in the illustration of the field effect in Fig. 3.13. In the case of the surface not under the high-vacuum conditions, the extent of the departure from the equilibrium condition in the bulk and expression in the form of surface potential is also influenced by the electrically charged molecules and ions originating from the ambient, which are adsorbed at the surface.

Surface potential of the bare surface cannot be measured in the way conducive with the needs of inline monitoring of surface-affecting processes, but the extent of the surface potential departure from electrical neutrality is reflected in the value of the surface charge associated with semiconductor surfaces discussed in the following section.

## 4.6 Surface Charge

As discussed earlier in this volume, density of the electric surface charge is an adequate representation of the broadly understood condition of semiconductor wafer surface. However, the discussion of the complex phenomena involving electric charge interactions with crystalline semiconductor surfaces during device manufacturing would be beyond the scope and the goals of this *Guide*. Therefore, in the following discussion, a simplified model is presented to justify the use of surface charge measurements in process monitoring applications considered in Chapters 7 and 8.

The origin of electric charge on crystalline semiconductor surfaces can be assumed to be twofold. First type of charge, denoted in Fig. 4.7(a) as $Q_1$, is related to disruptions of the crystallographic structure at the surface and in near-surface region, and is not directly dependent on the composition of the ambient to which the surface is exposed. Unless affected by thermal treatments altering its density, surface reconstruction, or other dedicated processes, it remains unchanged.

Another type of charge disrupting the equilibrium of semiconductor surfaces, denoted in Fig. 4.7(b) as $Q_2$, results from the interactions with an ambient to which the surface is exposed. As discussed in Section 3.2, these can be $H^+$ and $OH^-$ ions resulting from dissociation of water molecules contained in the moisture adsorbed at the surface or electrically

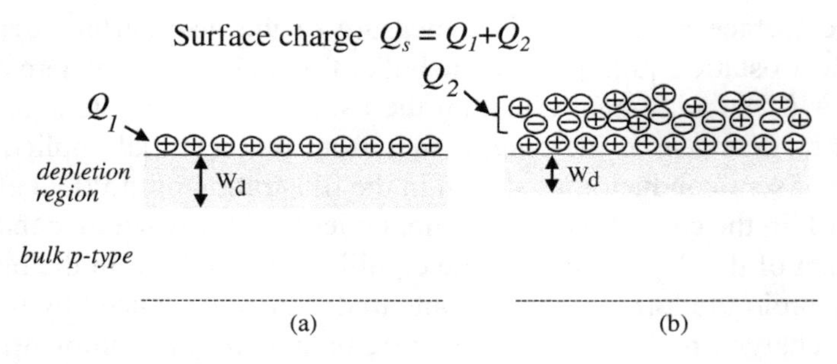


**Fig. 4.7.** Simplified representation of the surface charge $Q_s$ associated with (a) broken bonds and structural imperfections at the surface ($Q_1$) and (b) charges resulting from surface interactions with an ambient ($Q_2$).

active species originating from the organic molecules present in an ambient and adsorbed at the surface. Furthermore, contribution to the surface charge in this case may possibly come from the charges associated with the species accumulated on the surface as a result of cleaning or etching operations as well as from the process-related defects discussed earlier.

Total surface charge $Q_S$, shown in Fig. 4.7, is a cumulative surface charge that includes charge $Q_1$ as well as charges listed earlier, represented by $Q_2$. The density of surface charge varies, depending on the processes to which the surface is exposed and its value is an adequate representation of the performance of any given process to which wafer surface is subjected.

In short, while not qualitatively distinguishing between the nature of surface features, surface charge is a convenient quantitative representation of the broadly understood condition of semiconductor surfaces. As such, it can be used as a measure of its stability and reproducibility of its condition from run-to-run and from process-to-process during semiconductor device manufacturing. In Chapters 7 and 8, methods devised to measure surface charge and the ways of using such measurements in semiconductor device technology are described in more detail.

# Chapter 5

# Surface Effects in Semiconductor Devices

## 5.1 Introduction

Based on the earlier considerations regarding the characteristics of semiconductor surfaces, this chapter is concerned with functional semiconductor devices in which surface characteristics play a device performance affecting role. The goal is to identify devices in which surface-related effects influence their operation in the most evident fashion.

This chapter reviews the key classes of semiconductor devices distinguished in terms of their functions between electronic and photonic devices. Considering the lack of correlation between the characteristics of the surface of the wafer and the parts of the device system processed deep into its bulk, surface effects in micro-electromechanical semiconductor (MEMS) devices are not included in this survey.

The focus of discussion in this chapter is on electronic elements, which based on their operational fundamentals fall into the categories of unipolar and bipolar devices. With the role the condition of the surface plays in the operation of the former, a review in the remainder of this chapter is focused primarily on unipolar devices. An exception is an inclusion in the follow-up discussion of the comments regarding surface effects in solar cells, which by the nature of their operation belong to the bipolar class of semiconductor devices, but in which surface characteristics affect reflection and absorption of the sunlight and thus, impact their efficiency.

## 5.2 Surface Effects Depending on the Type of Device

Semiconductor materials serve useful purposes only when engineered into functional devices. The term *semiconductor device* is used here in reference to a material system (see Chapter 1), which is configured in a way that allows it to perform in a controlled fashion predetermined functions either electronic, or photonic, or electromechanical.

Electronic functions are performed by the devices the operation of which is based on the interactions between electric fields and electrons and holes acting as carriers of electric charge. In semiconductor devices such as those based on *p–n* junction, both electrons and holes are involved in their operation. Accordingly, corresponding devices are referred to as bipolar devices. In the case of unipolar devices, either electrons or holes control the operation of the device.

Figure 5.1 lists key electronic bipolar and unipolar devices and identifies those in which the condition of the surface of the substrate wafer influences their operation. As indicated, the condition of the surface of the substrate wafer plays a strong role in the case of unipolar devices and a relatively minor role in the case of bipolar devices.

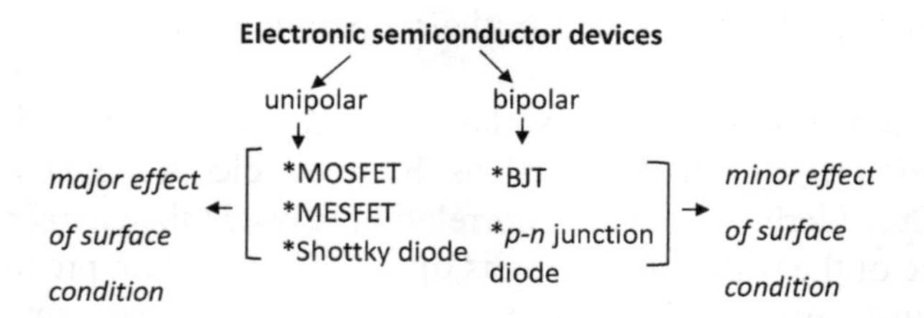


**Fig. 5.1.** Effect of surface condition on the operation of unipolar and bipolar semiconductor devices.

To justify conclusions regarding the role of the surface condition in the operation of the devices listed in Fig. 5.1, schematic cross-sections of the unipolar metal–oxide–semiconductor (MOSFET) in the configuration used across all its applications, both discreet and integrated, and the bipolar junction transistor (BJT) in the configuration used in integrated circuits are shown in Fig. 5.2. As depicted in Fig. 5.2(a), output drain current flows in the MOSFET in the direction parallel to the surface within an inversion layer forming channel in response to the voltage applied to the gate. Gate potential controls the current flow by changing the conductance of the channel through the field effect (see Fig. 3.13). Electrons moving in the channel between source (S) and drain (D) are subject to interactions with surface defects, such as scattering and recombination. In addition, surface roughness adversely affects the flow of electrons causing additional scattering and obstructing the flow of electrons in the channel (Ohmi *et al.*, 1991) in the ways discussed later in this chapter. Considering all the above, one may say that in the case of MOSFET, the surface is an active part of the device and its characteristics strongly affect the transistor's performance.

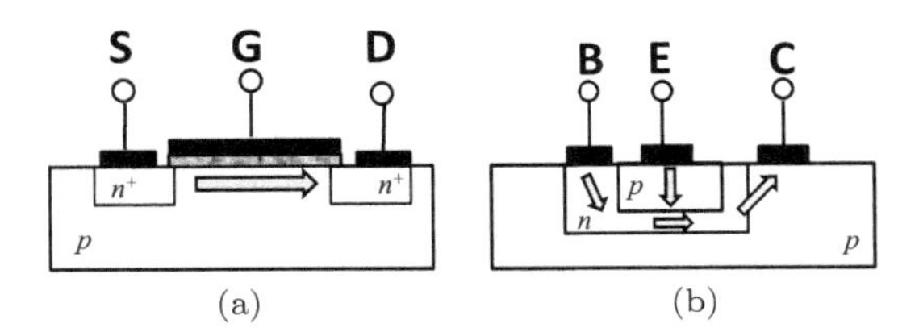


**Fig. 5.2.** Schematic representation of the currents flow in (a) metal–oxide–semiconductor field-effect transistor (MOSFET) and (b) *p–n–p* bipolar junction transistor (BJT).

In contrast, in the case of the integrated bipolar junction transistor shown schematically in Fig. 5.2(b), the surface of the semiconductor wafer has a limited effect on device operation. This is because the flow of the currents (emitter current $I_E$, base current $I_B$ and output collector current $I_C$) is controlled by *p–n* junctions and its operation under normal conditions is not affected by the condition of the surface. Only in the case of severe surface damage, there is a possibility of leakage current at the edge of the junctions at the surface affecting transistor operation. Also, because of the bipolar nature of this transistor, electron–hole recombination processes in the base region affect its operation rather than, as is the case in unipolar MOSFET, recombination of carriers carrying drain current with the surface and near-surface defects. The surface effects are even smaller in the case of the discrete bipolar transistors in which output collector current flows toward collector contact in the direction normal to the top and back surfaces of the wafer.

As it can be concluded from the above description of transistors' operation, the working of the unipolar field-effect transistors is expected to be affected by the condition of the surface of the semiconductor wafer.

At the same time, in the case of the bipolar junction transistor, characteristics of the surface of the substrate wafer, provided it doesn't feature major disturbances in terms of structural defects or contamination, have a relatively minor effect on the operation of the device.

In the remaining parts of this chapter, surface effects in the field-effect transistors are considered further.

## 5.3 Surface Effects in MESFET and JFET

In the case of unipolar field-effect transistors, the impact of surface characteristics on the transistor's output current varies depending on how the field effect is implemented. Two types of unipolar field-effect transistors, namely metal–semiconductor field-effect transistors (MESFETs) shown in Fig. 5.3(a) and junction field-effect transistor (JFET) schematically represented in Fig. 5.3(b) exemplify this notion. In the former case, the surface effects come into play, although in a different way and not as directly as in the case of the MOSFET considered in the following section. In the MESFET, the drain current flows in the channel conductance of which is controlled by the width *W* of the space-charge region formed at the metal–semiconductor interface (Fig. 5.3(a)). Control of the space-charge width depends on the condition of the surface in contact with the metal, and thus, surface processing prior to metal deposition plays a role in this case.

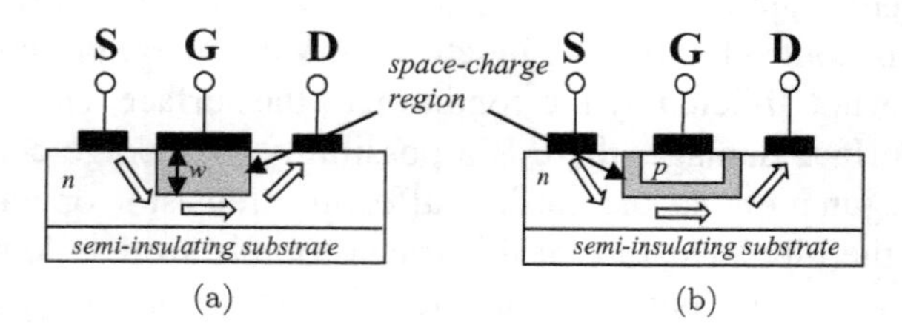


**Fig. 5.3.** Schematic cross-sections of (a) metal–semiconductor field-effect transistor (MESFET) and (b) junction field-effect transistor (JFET).

The surface effects, as understood in this discussion, have effectively no direct impact on device operation in the case of the junction FET (JFET). Here, the output drain current is controlled by the *p–n* junction space-charge region away from the surface (Fig. 5.3(b)).

In addition to those considered above, there are MOS, or MIS for metal–insulator–semiconductor, transistors operating on principles different from field-effect principles, yet in which condition of the surface also plays a performance-affecting role. This applies, for instance, to the metal–oxide–semiconductor transistors in the case when oxide is too thin to support the field effect and where tunneling across the oxide between metal and semiconductor controls device operation. The lateral MIS tunnel transistor (LMISTT) is an example of such a device (Ruzyllo, 1980). Here, the quantum effect of tunneling across ultrathin (2–5 nm) oxide controls transistor current and surface processing prior to thermal growth of ultrathin oxide plays a device performance-defining role.

As indicated in the above discussion, several types of semiconductor devices do not involve strong surface interactions affecting their operation. This observation applies primarily to the devices in which current flows in the direction normal to the surface between ohmic contacts on the front and back surfaces of the wafer. Also, in the case of multilayer device structures, where current flows horizontally within confined geometry of quantum well, for instance, such as in the case of high electron mobility transistors (HEMT), processing of the surface of the substrate wafer is not directly related to the device performance.

Based on the above reasoning, in the continuation of the discussion, the focus will be on MOSFET which features the most pronounced interplay between characteristics of the surface and operation of the device. In addition, MOSFET is a foundation of the complementary MOS (CMOS) technology which plays progress defining role in digital integrated circuits engineering and as such warrants special attention.

## 5.4 Surface Effects in MOSFET

The surface effects in the unipolar metal–oxide–semiconductor field-effect transistor (MOSFET) are used here to illustrate key issues surface processing in semiconductor device manufacturing needs to address. MOSFET was identified earlier as a device in which characteristics of the surface upon which the transistor structure is formed play a device performance-defining role (Wu *et al.*, 2001).

Electric charge carriers responsible for the output current of the transistor are moving from source to drain in the channel formed in the surface inversion layer which in advanced MOSFETs forming CMOS cells is a few nanometers deep. Therefore, charge carriers in the channel are subjected to interactions with surface features such as crystallographic defects, contaminants, and roughness, all of which can have an adverse effect on the density of the output current. In other words, both physical and chemical characteristics of the starting surface may affect channel conductance and reduce the flow of electric charge carriers responsible for device operation.

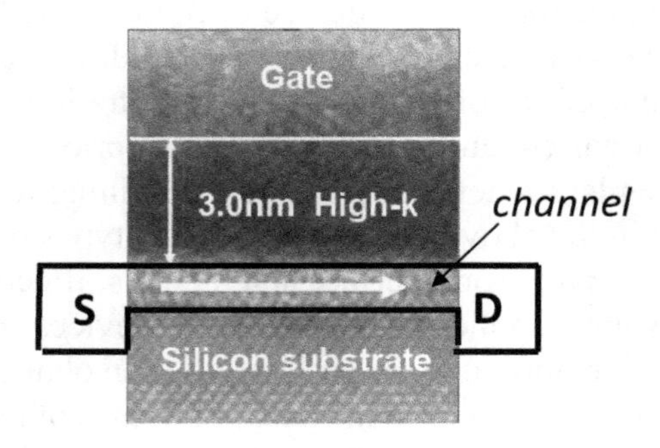


**Fig. 5.4.** Transmission electron microscope (TEM) image with superimposed parts of the MOSFET structure.

The transmission electron micrograph shown earlier in Fig. 1.7 and again in Fig. 5.4 with superimposed parts of the transistor, illustrates a disturbed crystallographic structure of silicon at the interface with high-*k* gate dielectric. As indicated in the diagram, charge carriers in the channel moving between source S and drain D are unavoidably subjected to interactions with defective crystal structures in the channel region. The measurable effect of surface condition shown in Fig. 5.4 on the transistor drive current manifests itself in the reduced charge carrier mobility in the channel caused by the scattering of electrons on the near-surface defects. Also, the recombination and trapping of moving charge carriers by interface defects have an adverse effect on the transistor current. Furthermore, interface defects may promote the tunneling of electrons across the interface barrier into the gate dielectric accounting for additional current losses.

The pre-gate dielectric deposition surface cleaning operations aimed at the removal of surface contaminants have no effect on crystallographic structure near the surface. As discussed in Section 3.4, the process of sacrificial thermal oxidation followed by oxide etching carried out prior to high-*k* dielectric atomic layer deposition (ALD) leads to a less disturbed silicon crystal structure at the surface, as compared to the one shown in Fig. 5.4. Due to its effectiveness, sacrificial oxidation is an often-pursued solution in silicon surface preparation procedures.

Modifications of MOSFET architecture toward 3D geometry, implemented to keep on increasing the density of digital integrated circuits while keeping transistors operational, call for continued changes in surface processing procedures in advanced MOS/CMOS technology (Sung *et al.*, 2020). The changes in question go beyond the challenges in the processing of FinFET structures where "fin" is typically defined by the etching of the substrate wafer and where the need to process vertical surfaces has brought about new solutions to transistor processing.

The MOSFET technology is encountering challenges different in nature as a result of the transistor architecture evolving from FinFET to GAA nanosheet FET (Fig. 5.5).

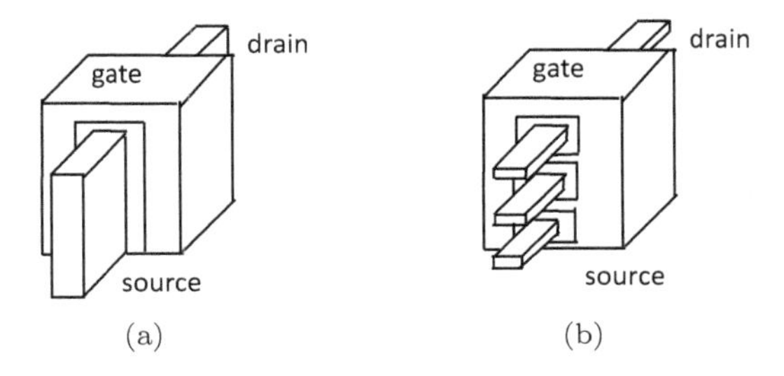


**Fig. 5.5.** (a) Schematic representation of FinFET and (b) gate-all-around (GAA) nanosheet FET.

With the transition of the FinFET architecture (Fig. 5.5(a)) to the gate-all-around (GAA) configuration with the channel in the form of a few nanometers thick nanosheets (Fig. 5.5(b)), the approach to pre-gate dielectric deposition surface processing needed to be adjusted accordingly. One of the reasons is that in the latter case, FET nanosheets acting as a channel do not originate from the substrate wafer, but are formed employing epitaxial deposition techniques or other techniques often involving recrystallization step, followed by the highly selective etching.

The methods used to achieve their formation and the sequence in which manufacturing steps are applied, make pre-gate dielectric deposition surface processing in the case of nanosheet-based FETs differ in several ways from those used in conventional FinFET technology. Particularly, in the case of complementary FET, or CFET in short, fabricated using stacked up $n$-type and $p$-type nanosheets which involve complex fabrication sequences and in which surface processing potentially causing pattern collapse should be avoided.

In addition, because of the variety of semiconductor materials that can be used to form nanosheets, including silicon, SiGe, binary and ternary semiconductor compounds or graphene, as well as different processes involved in their formation, it is not possible to identify surface processing procedures valid across all combinations of materials and process technologies involved in the manufacture of nanosheet-based gate-all-around transistors. It is, therefore, suggested that the readers interested in the state of the art in surface processing methodology in this area follow papers presented over the years during semiconductor surface cleaning and processing symposia (Symposia, 1989–2025, 1992–2025, 2002–2025).

As the discussion above demonstrated, using MOSFET and its derivatives as an example, the effect of the quality of the surface upon which some electronic devices are formed is of importance to the way they perform as well as to the manufacturing yield. The comments above also indicate that, with modifications in the MOSFET architecture, the approach to surface processing in the case of advanced logic and memory chips fabrication needs to evolve to respond to emerging needs.

## 5.5 Surface Effects in Solar Cells

Following on introductory discussion concerned with interactions of light with semiconductor surfaces in Section 3.3, the following considerations are focused on the requirements regarding surface characteristics specifically in the case of solar cells. In contrast to other semiconductor devices, the surface of semiconductor material in the case of solar cells is exposed to the environment, including sunlight, during device operation.

The common approach to the implementation of the photovoltaic effect converting sunlight into electricity is to form in semiconductor material a potential barrier using *p–n* junction which is located close to the surface. Solar light penetrating semiconductor material needs to get to the junction where the generation of electron–hole pairs takes place. Due to the electric field in the junction region, photo-generated carriers in motion are responsible for the current of the solar cell. For this sequence to work, solar light must reach the *p–n* junction which means that it must cross the surface of the cells exposed to sunlight without losses. To make it possible, the surface of the cell must be processed in a way that prevents the reflection of sunlight from the top surface of the cell (see Fig. 3.12).

If allowed, any reflection of the sunlight accounts for the losses resulting in the reduced efficiency of the cell.

The solution in this case is to texture the surface or in other words to roughen it in a controlled fashion (Kim *et al.*, 2020). In the case of the smooth surface, the probability of total reflection is high (Fig. 5.6(a)). In the case of the solar cell's surface that is textured in a deliberate way (Fig. 5.6(b)), interactions with the angled features of the surface eventually direct the incoming sunlight toward the *p–n* junction located at a certain distance from the surface.

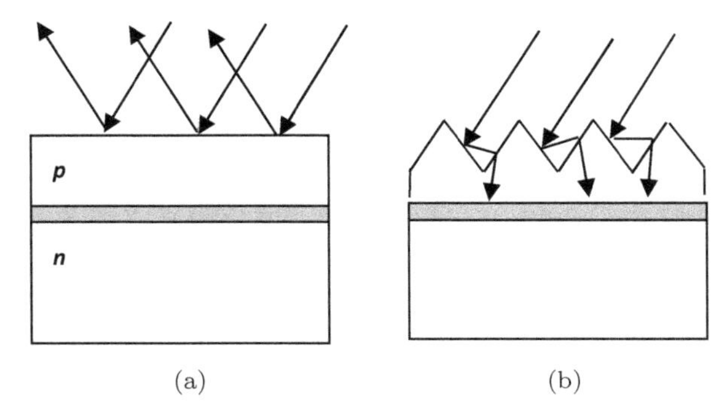


**Fig. 5.6.** (a) Solar cell with reflective surface and (b) the same cell with textured surface preventing reflection and enhancing absorption of sunlight.

Surface texturing shown in Fig. 5.6, when used in the case of multicrystalline solar cells, takes advantage of anisotropic etching of silicon, which features different etch rates along various crystallographic planes and produces the result, as shown in Fig. 5.7 (Drummond *et al.*, 2009).

The other method preventing reflection, one that can be used in the case of either crystalline or non-crystalline semiconductors, is concerned with the deposition of the thin layer of material transparent to sunlight acting as anti-reflecting coating (ARC) (Law *et al.*, 2023). When the surface of the solar cell is covered with ARC transparent to sunlight, which features the desired refractive index *n*, sunlight instead of being reflected passes through the ARC film and penetrates the silicon substrate. An added important function of the ARC film is to protect the surface of the cell against external influences including mechanical interactions of various natures.

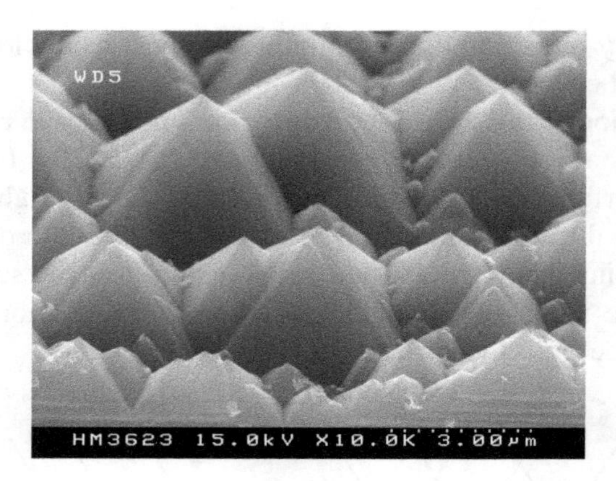


**Fig. 5.7.** Electron microscopy image of the surface of multicrystalline silicon texturized by anisotropic etching.

Yet another treatment of the surface, the goal of which is to increase the efficiency of silicon solar cells, involves covering its surface exposed to sunlight with a layer of material that broadens the absorption spectrum of the cell beyond the boundaries defined by the inherent characteristics of silicon. The procedure, referred to as down-shifting, can be implemented in various ways using various materials. As an example, the deposition of the film of cadmium selenide (CdSe) quantum dots on the surface of commercial silicon solar cells is effectively performing the down-shifting function. As demonstrated in Fig. 5.8, the layer of CdSe nanodots

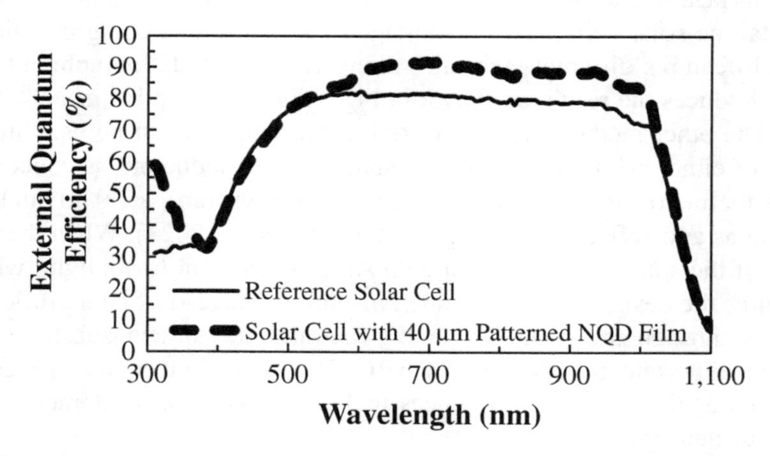


**Fig. 5.8.** External Quantum Efficiency of the commercial silicon solar cell with surface covered with Nanocrystalline Quantum Dots (NQD) downshifting film as a function of wavelength of light.

increases the efficiency of the cell in the UV portion of the sunlight spectrum featuring wavelength below 400 nm (Sabeeh *et al.*, 2019).

As it can be concluded from the discussion in this section, the effects in solar cells involving surface exposed to sunlight play a role in defining operation of these devices. Appropriate treatments of the surface minimize reflection of the sunlight, provide protection against external influences and may also be used to broaden the absorption spectrum of the cell resulting in the increased cell's efficiency in the UV portion of the sunlight spectrum.

# Chapter 6

# Surface Processing in Semiconductor Device Technology

## 6.1 Introduction

As emphasized in the discussion in previous chapters of this volume, conditions of the surfaces of semiconductor materials used to fabricate semiconductor devices play a role in the determination of operational characteristics as well as manufacturing yield of these devices, both electronic and photonic. In this chapter, operations performed on the bare semiconductor surfaces in the course of device fabrication with the goal of establishing surface characteristics conducive to the needs of the manufacture of functional semiconductor devices are considered. The discussion is focused on silicon which represents the most mature semiconductor surface processing technology among all semiconductor materials of interest. Not included in the overview in this chapter are surface finishing and cleaning steps which are part of the wafer fabrication procedures considered in Section 2.1.

Discussion in this chapter first considers surface contaminants encountered in the semiconductor device manufacturing environment, identifies their sources, and considers ways to control the level of contamination in this environment. In this context, process media including water, specialty chemicals, and gases are considered as possible sources of contamination. Subsequently, methods employed in the semiconductor industry to control the state of the processed surfaces through surface cleaning and conditioning are reviewed.

Surface cleaning is the process aimed at the removal of solid non-volatile contaminants from the surface and is not purposely geared toward the alteration of surface energy. Surface conditioning is understood here as a process, the purpose of which is to establish a predetermined chemical condition of the processed surface by means of the modifications of the surface energy through its activation and passivation processes. Various methods in both applications are considered, distinguishing between the methods carried out in the liquid-phase (wet processes) and the gas-phase (dry processes) media using silicon as a semiconductor material of reference (Ruzyllo, 2010).

Considerations of these topics are focused on the general principles underlying various surface treatments and do not attempt to bring to the attention of the readers emerging current solutions regarding broadly understood semiconductor surface engineering which undergoes frequent modifications. To stay current in this regard, it is one more time suggested that the readers follow the reports presented during topical symposia covering issues related to semiconductor surface processing (Symposia, 1989–2025, 1992–2025, 2002–2025).

The discussion in this part of the book is concluded with a brief overview of procedures used in device manufacturing involving semiconductor materials other than silicon and materials, most notably sapphire, used as a substrate upon which certain types of semiconductor devices are fabricated.

## 6.2 Surface Contaminants and Their Sources

Each industrial manufacturing process features requirements regarding the cleanliness of the environment in which any given fabrication procedures are carried out. Considering the nanoscale size of features created on the surface of the wafer, the number of operations performed on the wafer to produce advanced semiconductor devices including integrated circuits, as well as diverse media and elevated temperature processes involved, the needs of semiconductor device fabrication in terms of contamination control are the most stringent among all industrial endeavors.

### *Surface contaminants in semiconductor device fabrication*

Contaminants present in the semiconductor processing environment, which when allowed to interact with the surfaces of electrically neutral semiconductor wafers have a particularly deleterious effect on the manufacturing process and its outcome, and thus need to be controlled, are the following.

**Particles.** The contaminants which have an unquestionable adverse effect across all semiconductor process technologies are particles and particulates (groups of particles) adsorbed at the surface of the wafer being processed. Particles' contamination in semiconductor processing may originate from a multiplicity of sources including ambient air, liquid chemicals, water where the bacteria colonies end up acting as particles, and process gases. Furthermore, tools used to process semiconductor wafers as well as people working in the cleanrooms where semiconductor devices are manufactured generate significant amounts of particles.

In terms of makeup, particles encountered in a semiconductor device fabrication facility can be specks of dust, skin flakes, or colonies of bacteria. In addition, small chunks of semiconductor material being processed can contribute to particle contamination. The size of particles varies, depending on the source and its nature, from as large as 10 $\mu$m to as small as 100 nm and smaller in which case they could be difficult to detect. The problem is that while detection and control in the process environment of ultra-small particles remains to be a challenge, their effect on fabrication processes and manufacturing yield can still be destructive. For instance, even the smallest nanometer-sized particle on the silicon surface will obstruct pattern definition processes later in the device fabrication sequence. Particles contaminating bare silicon surface (Fig. 6.1(a)) will also adversely interfere with subsequent thermal oxidation, high-*k* dielectric or epitaxial deposition (Fig. 6.1(b)), contaminating at the same time the near-surface region of the wafer (Fig. 6.1(c)).

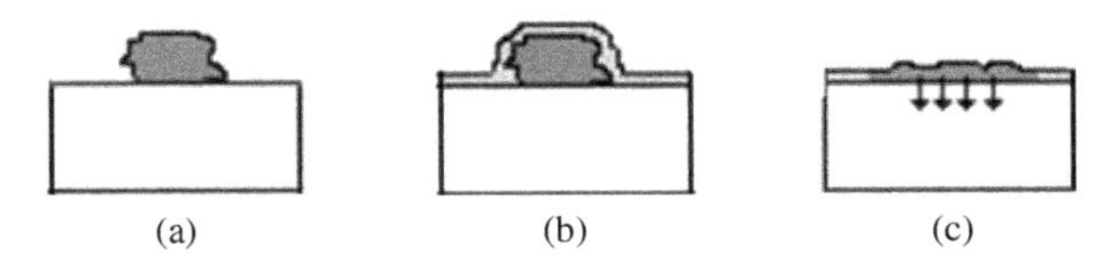


**Fig. 6.1.** (a) Particle on the bare silicon surface (b) interferes with subsequent thin-film deposition process and (c) may get decomposed during the follow-up processes involving elevated temperature, resulting in near-surface contamination of the substrate wafer.

Overall, the presence of particles in any semiconductor process environment is highly undesired and no efforts are spared to minimize their harmful effect. However, in light of the number of sources of particulate contaminants, total prevention of particle contamination of processed wafers is not possible. Therefore, the methods aimed at the removal of particles from the wafers' surfaces needed to be developed (Sahari *et al.*, 2009). These methods are an integral part of the routine wafer-cleaning operations discussed later in this chapter.

**Organic contaminants.** Volatile organic compounds are present in ambient air, including cleanroom air, in large amounts. They also originated from the processes exposing the surface of the wafer to organic compounds such as isopropyl alcohol (IPA) used in wafer drying operation. Furthermore, wafers are exposed to organic volatiles originating from storage and shipping containers (Saga and Hattori, 1996). Organic compounds, mostly in the form of hydrocarbons $C_xH_y$, readily adsorb on the semiconductor surface, as shown in Fig. 6.2, using silicon covered with native, spontaneously grown ultra-thin oxide $SiO_x$ as an example. If not controlled during prolonged wafer exposure to the ambient in the cleanroom or in the storage and shipping containers, organic compounds will interfere with the manufacturing of the functional devices. In combination with moisture, organic contaminants destabilize the characteristics of any semiconductor surface.

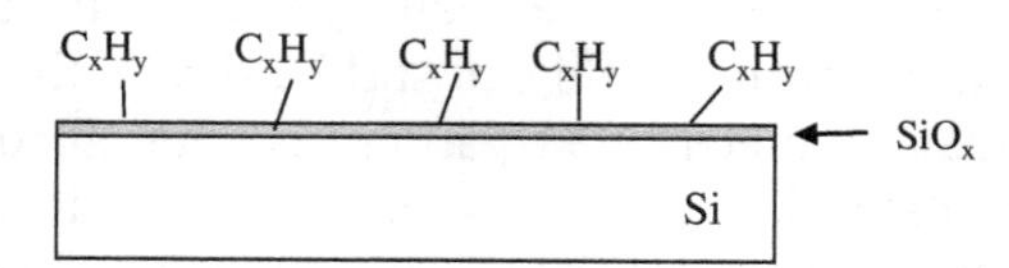


**Fig. 6.2.** Graphical representation of organic contaminants $C_xH_y$ on silicon surface covered with spontaneously grown native oxide $SiO_x$.

Semiconductor surfaces being prepared for the subsequent thin-film deposition step are particularly sensitive to organic contamination. This includes contact metallization processes by physical vapor deposition (PVD) as well as a range of additive processes implemented using atomic layer deposition (ALD) and chemical vapor deposition (CVD), including epitaxial deposition. The same considerations apply to solid organic residues remaining on the surface after the completion of the pattern definition processes ending with photoresist removal.

Under normal conditions, hydrocarbons are present in essentially any environment, including ambient air, liquids, and gases. As such, organic compounds cannot be eliminated from the ambient air, including ultra-clean cleanroom air, and their adsorption on the processed surfaces cannot be entirely prevented. Therefore, operations aimed specifically at the removal of organic contaminants from the processed semiconductor surfaces are an integral part of the cleaning procedures employed in semiconductor device manufacturing.

**Metallic contaminants.** Metallic contaminants originate primarily from liquid chemicals and water as well as from metal parts of the process tools, tubing, and plumbing hardware supporting the operation of the semiconductor manufacturing facility. They may also contaminate the processed surface as a result of physical contact with metal parts during wafer handling operations.

The deleterious effect of metallic contaminants extends across the broad range of characteristics of semiconductor devices, particularly those which require elevated temperature processing during manufacturing such as thermal oxidation or diffusion. Temperature activates metallic contaminants on the surface and converts them into defects having strongly adverse effects on the characteristics of the near-surface region of the semiconductor substrate.

The most common in semiconductor process environment metallic contaminants are iron (Fe), aluminum (Al), copper (Cu), nickel (Ni), as well as ionic metals, such as sodium (Na) and calcium (Ca). If left on the surface of the wafer that is subjected to high-temperature treatment, metallic contaminants may cause severe damage not only to the surface of the wafer but also, depending on the segregation coefficient in silicon, to its bulk characteristics (Fig. 6.3).

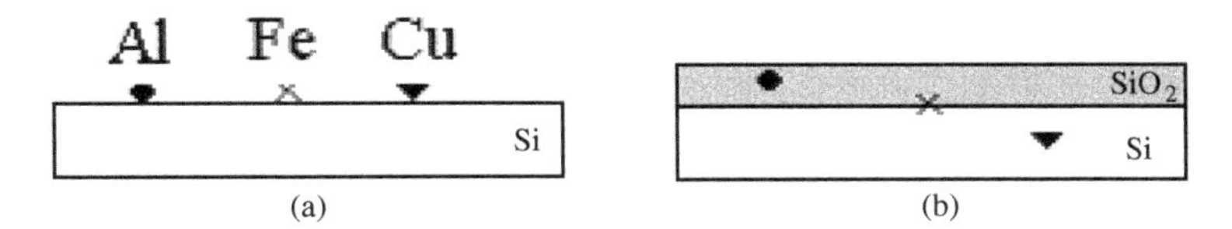


**Fig. 6.3.** (a) Common in semiconductor process environment metallic contaminants on Si surface and (b) segregation coefficient-driven distribution of the same metallic species in Si–$SiO_2$ after thermal oxidation.

As can be seen in Fig. 6.3(b), each metallic contaminant of interest interacts with silicon subjected to high-temperature thermal oxidation differently. The segregation coefficient of iron in silicon is such that during oxidation, iron atoms remain in the near-surface region of the silicon wafer forming electrically active defects altering the transport of charge carriers in the affected region. Copper, in turn, featuring a high diffusion coefficient in silicon during thermal oxidation diffuses deep into the bulk of silicon wafers where it forms recombination centers affecting current flow characteristics. The opposite happens during thermal oxidation to surface-contaminating aluminum. Segregation coefficients of Al in silicon and in growing $SiO_2$ are such that aluminum atoms remain near the top surface of the growing oxide (Fig. 6.3(b)). The presence of aluminum-related centers in the oxide adversely affects the oxide's breakdown characteristics.

For the reasons identified above, and despite the decreasing level of metallic contamination of process chemicals and water used in state-of-the-art semiconductor manufacturing facilities, in some wafer cleaning sequences applied prior to high-temperature processes in particular, steps aimed specifically at the metallic contaminants' removal are required. In some other cases, including semiconductor surfaces prepared prior to metal contact deposition, requirements regarding the level of metallic contamination are less stringent.

**Moisture.** While not strictly speaking a contaminant, moisture on the surface of the wafer may not only alter the chemical conditions of the surface and interfere with subsequent processes performed on the wafer, but it may also affect the results of surface characterization measurements to which semiconductor surfaces are subjected. For these reasons, it is essential that the moisture on the surface of the wafer is controlled by controlling the moisture content in the ambient to which the wafer is exposed as well as the time of exposure to this ambient. Moisture on the surface of the wafer promotes chemical reactions on the surface, has a destabilizing effect on the surface chemistry, as well as plays a role in altering the kinetics of various water-content-sensitive processes. For the sake of illustration of the nature of this last effect, Fig. 6.4 shows the dependence of the thermal oxide etch rate in anhydrous hydrofluoric acid (AHF) mixed with the vapor of methanol ($CH_3OH$) into what is known as AHF/MeOh mixture (see the discussion in the following section) as a function of wafer storage time after it has been thermally oxidized (Staffa and Ruzyllo, 1996). The increasing etch rate seen in this figure is due to the increasing amount of moisture absorbed by the oxide during storage, promoting oxide etching.

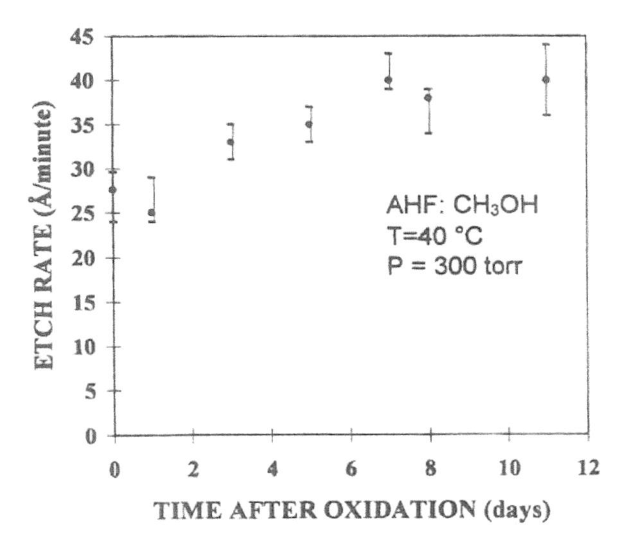


**Fig. 6.4.** Change of $SiO_2$ oxide etch rate in AHF/MeOH vapor as a function of wafer storage time in ambient air after thermal oxidation.

**Surface residues.** The products of the processes performed on a semiconductor surface such as, for instance, etching, may also be considered as surface contaminants which need to be removed by dedicated treatments. Moreover, spontaneous surface reactions specific to any given semiconductor material tend to alter the chemistry of its surface in an undesired fashion. In the case of silicon, due to its high affinity to oxygen, silicon surface treated in water or exposed to ambient air spontaneously grows an ultra-thin layer of its native oxide $SiO_x$ featuring uncontrolled composition (Fig. 6.2). The presence of residual native oxide on the silicon surface is bound to affect the outcome of various subsequently applied processes such as the epitaxial deposition or formation of the ohmic contacts. The oxide-like film present on the surface also interferes with various characterization measurements performed on the surface. Furthermore, it may also disrupt cleaning processes aimed at the removal of particles and other contaminants discussed above. For this reason, control of the native oxide on the surface of a silicon wafer is part of the surface cleaning operations applied in silicon device technology.

## *Sources of contamination in semiconductor device fabrication environment*

As indicated in the above discussion, surface contaminants listed originate from the ambient in which semiconductor wafers are processed during device fabrication and are not concerned with contaminants in the bulk of the wafer away from its surface. Those which could have penetrated the wafer during high-energy interactions such as ion implantation or ion etching are beyond the scope of this discussion.

The elements of the semiconductor device process technology which may act as the sources of surface contamination, and thus need to be monitored and controlled, are considered below.

**Water.** Water is a medium of special importance in semiconductor manufacturing where it is used in large amounts. This is despite the efforts to limit its use in the semiconductor industry because of the high cost involved and environmental concerns.

During semiconductor device fabrication, water serves three main purposes. First, it is an agent which is used to establish the desired composition of the chemical solutions used to process wafers. Second, water is used to stop chemical reactions by overflowing reactive chemistries to which semiconductor wafers are exposed. Finally, water is an agent used to remove products of chemical reactions from the surface through the process of rinsing.

As an extensively applied medium, water, if not properly processed, can be a source of any of the surface contaminants considered above. For this reason, in semiconductor device processing, water that is particle-free and chemically pure is used. The measure of water purity is its electrical resistivity. Water, not exposed to any ambient, is considered to be pure at 25°C when its resistivity reaches 18.2 MΩ-cm, at which point it is controlled solely by $H^+$ and $OH^-$ ions. Water used in high-end semiconductor manufacturing should feature resistivity of about 18 MΩ-cm to ensure adequate performance of various surface treatments requiring interactions with liquids.

To accomplish this level of chemical purity, filtrated water is subjected, employing the process of reverse osmosis to deionization (Fig. 6.5). The resulting deionized water, DI water in short, is a standard in any semiconductor research laboratory and manufacturing facility.

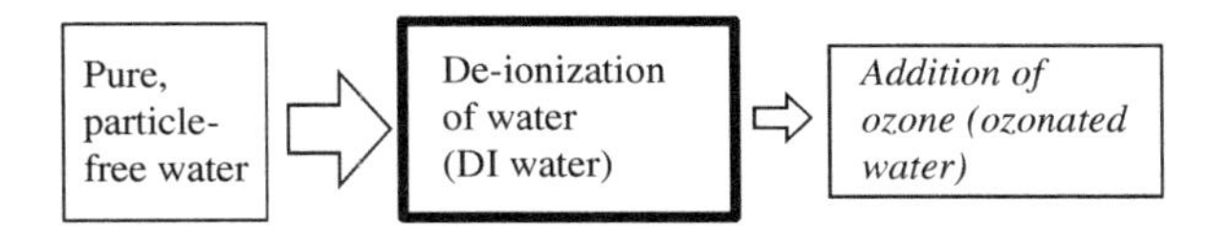


**Fig. 6.5.** The sequence applied in the preparation of water for use in semiconductor device technology.

To enhance water's ability to control undesired organic contamination, including colonies of bacteria which may develop into particles adhering to the processed surfaces, controlled amounts of ozone acting as a strong oxidizing agent are dissolved in DI water (Fig. 6.5). The resulting ozonated water is used in the semiconductor industry in the manufacture of advanced elements.

**Chemicals.** Large quantities of liquid specialty chemicals are consumed in semiconductor processes involving surface cleaning operations using liquid chemicals and water (wet cleaning) as well as in material removal by means of etching in the liquid phase. The choices of liquid chemistries used depend on the chemical composition of the processed materials and may vary between various solids processed in the course of semiconductor device fabrication keeping in mind the effect surface treatments may have on the follow-up processes (Ruzyllo, Rohr *et al.*, 1999).

Just like in the case of water, an issue of importance in wet process technology in semiconductor manufacturing is concerned with the purity of chemicals used. Only the highest purity chemicals are suitable for surface processing in high-end semiconductor manufacturing as any contamination of liquid chemicals with particles and trace metallic impurities will have an adverse effect on the performance of the manufacturing process and eventually on the performance of fabricated devices.

In addition to the possible contamination of those originally delivered, reactive chemicals used in semiconductor manufacturing can also be contaminated because of the malfunction of the complex chemicals delivery systems in the device fabrication facility. For instance, erosion of the metal elements within the delivery system can be an often-encountered source of contaminants. In large semiconductor manufacturing facilities, to reduce the chances of contamination of liquid chemicals and to improve safety associated with handling of large volumes of highly reactive liquids, a procedure of point-of-use chemical generation is implemented. By following this procedure, aqueous chemicals needed are generated by *in situ* (point-of-use) mixing of ultra-pure DI water with gaseous chemicals.

**Gases.** Essentially any semiconductor device manufacturing process requires gases which either provide a controlled process environment or are chemically or physically engaged in the process. Similar to liquids, also in the case of gases used in semiconductor device processing, their purity is of major concern. Only gases featuring the highest purity expressed in percent, e.g. >99.9995% and higher, can be used in advanced semiconductor device manufacturing. In some applications, gases as pure as 99.99999% are required.

The above requirement applies to both inert gases such as nitrogen, $N_2$, and argon, Ar, as well as process gases, also referred to as specialty gases, which interact chemically with other gases or with solids surfaces which are exposed to them. An important characteristic of nitrogen from the practical application point of view is its relatively high boiling point which allows the conversion of gaseous nitrogen into liquid nitrogen (LN) on the industrial scale. Under normal atmospheric pressure, nitrogen can exist as a liquid up to the temperature of 77.2 K (−196°C). From the surface contamination perspective, nitrogen gas originating from liquid nitrogen is preferred over nitrogen delivered in pressurized tanks.

**Process tools.** In addition to the sources of contamination considered above, processes used in device manufacture on their own often generate contaminants which may end up being adsorbed at the wafer surface. The schematic diagram in Fig. 6.6 represents a fictitious reactor of the type used in semiconductor device fabrication identifying possible contamination-generating parts and interactions.

For instance, particulate contamination could be a product of some etch processes. Besides, essentially all tools used in semiconductor manufacturing involve robotic wafer handlers, featuring moving parts. Unless designed specifically to cope with friction-resulting contamination, moving parts within the tool may generate large amounts of particles of various nature. Furthermore, materials used to construct process chambers designed to carry out processes employing aggressive chemical reactions must maintain their chemical integrity at elevated temperatures. As a matter of fact, elevated temperature itself may cause outgazing of contaminants from the process chamber even without any chemical reactions involved. The same applies to materials used to fabricate cassettes, boats, containers, and other parts used to handle wafers during processing.

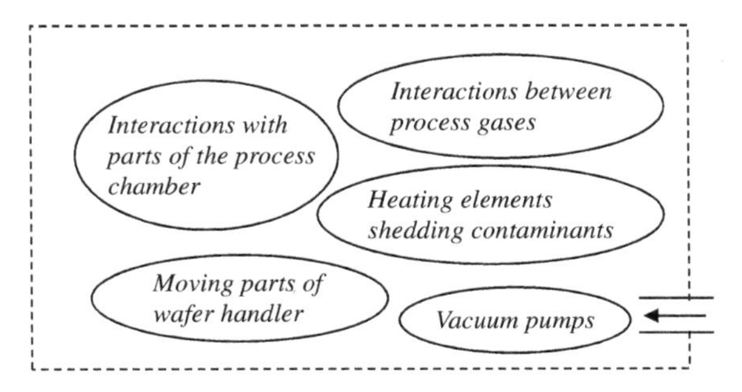


**Fig. 6.6.** Schematic representation of contaminants generating interactions inside the reduced pressure reactor used in semiconductor manufacturing.

Also of importance are utensils coming into physical contact with the wafer surface. In light of the possible contamination with metallic species and physical damage to the surface, metal tweezers are not used for handling semiconductor wafers. Instead, Teflon tweezers or vacuum-based handling tools are used.

**Cleanroom environment.** To prevent contamination of the wafer surface during device fabrication, not only contamination-free process media but also ultra-clean environment in which manufacturing processes are taking place must be assured. A well-known concept of a cleanroom refers to the enclosed environment in which semiconductor manufacturing processes must be carried out to prevent particle contamination of the processed surfaces. Moreover, cleanrooms are designed to assure strict control over temperature, moisture content in the ambient air, and static electricity, all of which may have an adverse effect on the condition of the surfaces of the processed wafers.

The most common measure used to define the particle contamination level in the cleanroom air divides cleanrooms into classes from class 1 to class 100000, depending on the number of particles of a given size per cubic foot of air which in turn depends on the type of filters used. Accordingly, representing the clean air, class 1 allows one 0.5 $\mu$m particle per cubic foot of air, and no particles larger than 0.5 $\mu$m. This level of particle control is needed in the manufacture of integrated circuits featuring low-nanometer geometries and requires ultra-low penetration air (ULPA) filters while for the class 100 cleanroom and above, high-efficiency particulate air (HEPA) filters are being used. Both ULPA and HEPA filters work in conjunction with a laminar flow of the air recirculated through the filters as opposed to turbulent flow which in terms of particle handling is more disruptive.

**Shipping and storage containers.** An issue of importance in semiconductor research and industry concerns containers used to store wafers between processing steps in the device manufacturing sequence, as well as those used to ship wafers. Commonly used containers tend to outgas organic compounds which end up being adsorbed on the surfaces of the shipped wafers. These effects must be accounted for by dedicated cleaning operations performed on the wafers before starting the manufacturing process.

**Additional consideration.** Following on the review of surface contaminants and contamination control in the semiconductor process environment, it needs to be emphasized that different types of contaminants may interact differently with different materials under the same process conditions. Each type of contaminant has the potential for interfering with the fabrication process as well as adversely affecting the performance of the final device, but depending on the process and semiconductor material, the effects could be different. For instance, surface contamination with metallic species is a major issue in case bare silicon surface is subjected to high-temperature thermal oxidation of silicon or epitaxial deposition but is less of a concern in the case when the same surface metallic contaminants interact with the silicon surface prior to contact metallization. Therefore, the discussion of the role of various surface contaminants needs to be considered in the context of specific materials and manufacturing processes.

## 6.3 Cleaning of Silicon Surfaces

As it can be concluded from the discussion in the previous section, total prevention of surface contamination during semiconductor device fabrication is for all practical purposes not possible. This is because in spite of using a semiconductor device technology with the highest purity water, chemicals, and gases, and carrying out device manufacturing in an ultra-clean environment, the tools used in semiconductor processing and operations they perform by themselves are often the source of contamination (Fig. 6.6). Therefore, elaborate methods of semiconductor surface cleaning need to be applied in semiconductor device manufacturing to achieve contamination-free wafer surfaces. Otherwise, either manufacturing yield or adequate device performance cannot be realized.

In this section, general principles upon which semiconductor cleaning methods are based are briefly reviewed, focusing on the cleaning procedures in the case of bare silicon surfaces. The follow-up considerations do not attempt to track current developments in the evolving field of semiconductor cleaning technology. For information of this nature, readers are referred to the reports from the symposia devoted to semiconductor surface cleaning science and technology and other sources devoted to this topic (Symposia, 1989–2025; Kern, 1990, 1993; Hattori, 1998; Reinhardt and Reidy, 2011).

The goal of the cleaning operations in semiconductor device manufacturing is to produce a surface that is clear of any chemical elements or molecules other than the host material, which have reached the surface in an uncontrolled fashion, as well as particles of any kind. Cleaning action itself can be implemented in various ways including methods schematically illustrated in Fig. 6.7. The most common one involves the chemical reaction between the reactant in the cleaning ambient and the contaminant on the surface (Fig. 6.7(a)). It is often accompanied by physical interactions between the cleaning ambient and the surface which are meant to facilitate removal from the surface of any remaining alien species (Fig. 6.7(b)). Another approach may rely on the momentum transfer between high kinetic energy species directed toward the surface and the contaminants such as particles (Fig. 6.7(c)). Yet another solution is to use light to cause desorption of volatile contaminants from the surface (Fig. 6.7(d)).

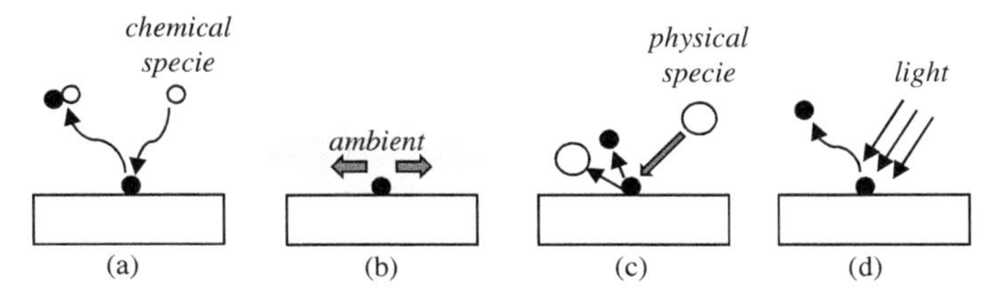


**Fig. 6.7.** Implementation of wafer cleaning processes in semiconductor device manufacturing by means of (a) chemical reaction, (b) megasonic agitation, (c) physical interactions, and (d) illumination-driven volatilization of contaminants.

In practice, cleaning methods are differentiated based on the ambient in which cleaning action is implemented. Generally adopted classification of cleaning operations in semiconductor manufacturing distinguishes between wet cleaning using liquid-phase chemicals and water, and dry cleaning carrying out surface cleaning action in the gas-phase or in a vacuum. An alternative technique discussed later in this section, known as supercritical cleaning, is used to clean patterned structures which may or may not involve interactions with as-processed bare semiconductor surfaces.

**Liquid-phase (wet) cleaning.** Wet cleaning is a mainstream cleaning technology in semiconductor device manufacturing. The goal of the follow-up discussion is to introduce general rules involved in wet cleaning procedures in semiconductor device manufacture using silicon wafers as an example. In light of the improved contamination control in semiconductor device manufacturing, cleaning technology is evolving according to the needs.

Wet cleans use combinations of acids, solvents, surfactants, and deionized water (DIW) to dissolve, oxidize, etch, and scrub contaminants from the wafer surface. In other words, wet cleaning involves processes in which contaminants are removed via a selective chemical reaction taking place in the liquid phase which causes either their dissolution in the solvent or conversion into soluble compounds (Fig. 6.7(a)). Typically, the process is enhanced by physical interaction with a cleaning solution such as megasonic agitation producing sonic waves along the cleaned surface of the wafer (Fig. 6.7(b)).

Wet cleaning recipes vary depending on the semiconductor material subjected to cleaning and the extent of surface contamination. In the case of silicon, they draw from the pool of chemical mixtures that were first proposed as the so-called RCA clean (Kern and Puotinen, 1970). Since then, it has been modified in various ways toward simplification of cleaning sequences and savings on resources such as high-purity chemicals and deionized water. In more recent years, it is also referred to as a Standard Clean, SC in short.

The general principle in terms of cleaning chemistries used and the order in which they are applied remain mostly the same. Table 6.1 (Heyns *et al.*, 1999) lists the chemical solutions commonly used to remove from silicon the surface contaminants discussed earlier in this chapter using broadly adopted semiconductor terminology acronyms. Among them, a solution denoted as APM (ammonium peroxide mixture), also known as

**Table 6.1.** Wet cleaning recipes commonly used in silicon technology.

| **Particles** | **Organics** | **Metallics** | **Native oxide** |
|---|---|---|---|
| APM (SC-1) | SPM | SPM | HF:$H_2O$ |
| | APM | HPM (SC-2) | |
| | | DHF | |

Standard Clean-1, or SC-1, represents ammonium hydroxide ($NH_4OH$): hydrogen peroxide ($H_2O_2$):deionized water ($H_2O$) mixture applied at 40–70°C to remove from silicon surface particles, heavy organics, and trace metallic contaminants possibly remaining on the surface after photoresist stripping. When there is a need to remove from the surface organic contaminants, a sequence involving sulfuric acid ($H_2SO_4$):hydrogen peroxide ($H_2O_2$):deionized water ($H_2O$) mixture at 100–130°C, known as SPM (sulfuric peroxide mixture) or "piranha" clean, followed by APM exposure, is used.

In the case of metallic contaminants removal from the silicon surface (Table 6.1), the SPM exposure is followed by treatment in HPM (hydrochloric peroxide mixture) solution composed of hydrochloric acid ($H_2SO_4$):hydrogen peroxide ($H_2O_2$):deionized water ($H_2O$) mixture, also known as Standard Clean-2, or SC-2. In most of the applications, the SPM–HPM sequence needs to be followed by the removal of the spontaneously grown native silicon oxide using diluted hydrofluoric acid (DHF). This last step, however, is of limited use in the case of processing of silicon surface immediately before the deposition step requiring an entirely oxide-free Si surface. This is because the DHF etching step is typically followed by DI water rinse causing regrowth of residual native chemical oxide on the silicon surface.

The above sequences are resource- and time-consuming and in practice are simplified in a variety of ways. For instance, there are applications in which a sequence involving just one cleaning step followed by DI water rinse and drying is sufficient. An example of such a simplified, yet effective sequence is the process known as "IMEC-clean" comprising an ozonated water exposure followed (without a rinse) by a treatment in a weak solution of HF:HCl:$H_2O$ (e.g. 0.025:1:5) and completed with a rinse and dry cycle.

Besides the chemical composition of the wet cleaning solutions, the way those solutions are delivered to the wafer for the purpose of surface cleaning is an important consideration. Wafers can be subjected to cleaning solutions one at a time (single wafer cleaning) or in batches in which the number of wafers depends on the wafer size (batch cleaning). Furthermore, the direct contact between the cleaning solution and the wafer's surface subjected to cleaning can be established in various ways. The most common involves the immersion of wafers (or a wafer) in a cleaning bath (Fig. 6.8(a)).

This method of immersion cleaning assures uniform exposure of both surfaces of the wafer to cleaning chemistries and is compatible with megasonic agitation which, by generating sonic waves propagating in the direction parallel to the wafer surface, increases the efficiency of cleaning, especially particle removal. It also allows precise control of the temperature of the bath and assures uniform rinsing, completing the cleaning sequence. A downside of this approach is that it requires significant amounts of process chemicals and deionized water, and may cause a collapse of fine patterns created on the surface.

Immersion cleaning is implemented in wet benches which include several cleaning/rinsing tanks. Cassettes with wafers are automatically moved in the desired sequences from tank to tank, as illustrated in Fig. 6.8(a). The complexity of the wet benches in terms of the number of tanks used for wafer cleaning depends on the cleaning sequence. In either case, tanks dedicated specifically for the DI water rinse operation are inherent parts of the cleaning apparatus commonly referred to as the wet bench. Immersion cleaning is also implemented in a single-wafer cleaning mode.

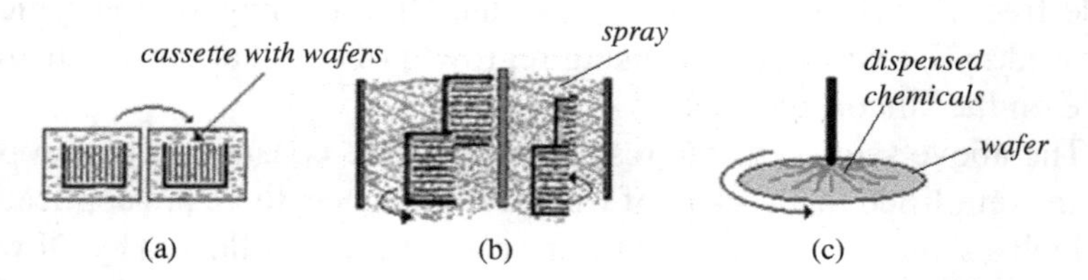


**Fig. 6.8.** Implementation of wet cleaning using (a) immersion cleaning, (b) spray cleaning, and (c) single-wafer spin cleaning.

Alternative approaches to wet cleaning in semiconductor manufacturing include centrifugal spray cleaning in which cassettes with wafers are rotated in the spray of cleaning chemicals, or rinsing water, dispensed from the nozzles (Fig. 6.8(b)). This technique offers savings on chemicals and DI water, but its uniformity in select high-end applications may not be sufficient. Yet another approach known as spin cleaning is a single-wafer process typically used in the case of large-diameter wafers (Fig. 6.8(c)). In this technique, chemicals, or rinsing water, are dispensed onto the rotated wafer. To increase the efficiency of wafer cleaning, a higher-pressure spray, or jet spray, can be employed and the dispensing nozzle can be moved along the diameter of the rotating wafer.

The performance of each technique may vary from application to application and the choice is based on the balance between the required performance and the cost of chemicals and water used. The efficiency of wet wafer cleaning technology is proven and well documented (Symposia, 1989–2025, 1992–2025, 2002–2025).

All in all, wet cleaning, despite its shortcomings related to the cost of consumables, is a mainstream technique in semiconductor device manufacturing due to its ability to remove all kinds of contaminants of interest and its overall efficiency. At the same time, however, efforts aimed at the reduction of the number of cleaning operations performed on the wafers during semiconductor device fabrication represent a clearly identifiable trend.

An inherent part of any process carried out in the liquid phase and ending with water rinse includes wafer drying operations. Water left on the surface to evaporate attracts particles which leave traces on the surface known as watermarks. Thus, the process of removing water from the surface, or drying, is a step which determines the condition of the surface of the wafer following any wet treatment.

A somewhat rudimentary wafer drying technique is based on spin drying in which water is removed from the surface by centrifugal forces during the fast spinning of the wafer in clean air. A wafer drying method of choice in demanding commercial processes involving nanometer-scale device features is using isopropyl alcohol (IPA) ambient and exploiting the Marangoni effect. Known as IPA drying, or Marangoni drying, the method uses the difference between the surface tension of the solid in contact with deionized water and IPA vapor. Surface drying is accomplished by pulling a sample, for instance a semiconductor wafer, out of the water into the volume of IPA vapor and nitrogen which results in the effective removal of the water molecules from the wafer surface (Ruzyllo, 2020).

**Supercritical cleaning.** While remaining a cornerstone of semiconductor surface processing since its introduction, liquid-phase cleaning features shortcomings which include surface tension-related inability of aqueous cleaning chemistries to penetrate high-aspect-ratio nano-patterns.

The solution to the above limitations of liquid-phase processing is provided by supercritical cleaning which instead of water uses supercritical fluid (SCF) as the carriers of cleaning agents. Supercritical fluids represent a state of matter into which gases and liquids can be converted at elevated temperatures and high pressure and display characteristics of both liquids and gases. Most notably, supercritical fluid lacks surface tension while interacting with solid surfaces and hence, can penetrate high-aspect-ratio geometrical features patterned into semiconductor materials (Fig. 6.9).

Moreover, SCF features low viscosity and, like a liquid, easily dissolves large quantities of liquid chemical compounds carrying them into very tight geometrical features as shown in Fig. 6.9. The gas of choice for supercritical fluid cleaning applications is carbon dioxide ($CO_2$) known as supercritical $CO_2$ or SCCO2. A critical point at which $CO_2$ transitions from gas to supercritical fluid is 31°C at the pressure of 31 atm.

Industrial use of supercritical $CO_2$ is not bringing about any major environmental concerns because carbon dioxide in SCCO2 can be recovered and reused.

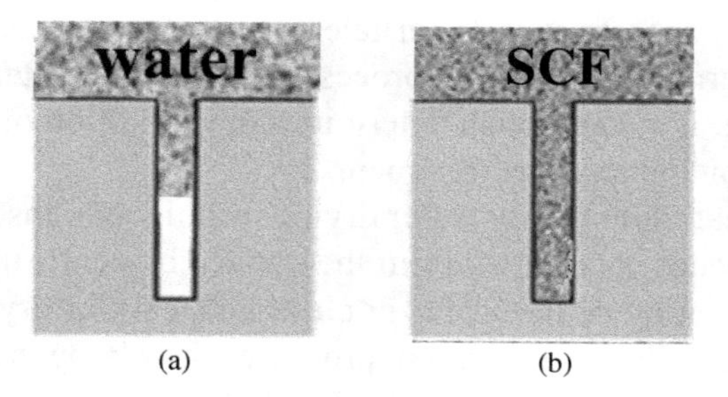


**Fig. 6.9.** Unlike (a) water-based liquid, supercritical fluid (SCF) (b) penetrates tight geometrical features.

**Gas-phase (dry) cleaning.** In addition to the constraints listed above, possible pattern collapse during wet cleaning and rinsing steps, as well as potential challenges regarding the integration of wet surface treatments with subsequent, typically vacuum-based, thin film deposition steps may also be of concern in some applications. The high cost of chemicals and ultra-pure DI water used in wet cleaning operations as well as environmental issues related to waste management add to the concerns regarding liquid-phase cleaning technology.

In light of the limitations faced by wet cleaning methods in some applications, gas-phase (dry), often implemented in the vapor phase, processing methods were proposed and evaluated (Ruzyllo, 1988; Ruzyllo *et al.*, 1989; Beck *et al.*, 1994; Deal *et al.*, 1990) and the key challenges were identified (Ruzyllo, 1990, 1993; Butterbaugh and Muscat, 2007). In addition to surface reactions driven by chemical reactivity of reagents, gas-phase surface processing can be enhanced by light, both visible and UV (Ito *et al.*, 1990), laser beam, as well as by plasma (Yew and Reif, 1990; Yuh *et al.*, 1998) or energetic species accelerated toward the surface and carrying energy which can be used to enforce directionality (anisotropy) of gaseous ambient interactions with processed surface (Staffa *et al.*, 1999). This contrasts with wet processes which by relying on the chemical reactions cannot be enhanced toward compatibility with applications involving non-planar, deeply etched semiconductor surfaces considered later in this chapter.

Dry and wet processes are in some situations similar in terms of the ability of each to support chemical reactions designed to accomplish specific process goals. For instance, wet processes using HF:water solution can be replaced with anhydrous HF:alcoholic solvent vapor (Torek, Ruzyllo *et al.*, 1995) in certain native oxide etching operations facilitating their integration with the next process in the device fabrication sequence carried out in the gas phase (Ruzyllo, Frystak *et al.*, 1990; Ma *et al.*, 1995).

The follow-up discussion provides a concise assessment of the performance of gas-phase methods in surface cleaning applications in the case of contaminants commonly encountered in the semiconductor process environment and discussed earlier in this chapter. Examples of gas-phase methods considered for various semiconductor surface cleaning applications are summarized in Table 6.2.

In terms of particle removal from semiconductor surfaces using dry methods, the use of a cryogenic aerosol comprising high-velocity frozen particles of inert gas such as nitrogen and argon (Saito *et al.*, 2004; Banerjee, 2015) as well as carbon dioxide $CO_2$ impinging on the surface and dislodging particles through the momentum transfer is a solution. Alternative gas-phase approaches involve particulate contamination

**Table 6.2.** Examples of possible gas-phase surface cleaning and conditioning methods.

| Particles | Organics | Metallics | Native oxide |
|---|---|---|---|
| *Cryogenic aerosol Ar/$N_2$<br>*$CO_2$ snow | *$O_2$/Air anneal<br>*UV/ozone<br>*Lamp (thermal) cleaning | *Anneal in Cl-based chem.<br>*UV/$Cl_2$ or UV/$SiCl_4$ | *$H_2$ anneal<br>*AHF/$H_2O$<br>*AHF/alcoholic solvent |

removal using laser-assisted methods (Lee *et al.*, 1993). In all these cases, the efficiency of removing low nanometer-sized particles from the bare surfaces of large-diameter silicon wafers could be a challenge. As a result, the usefulness of gas-phase methods in surface particle control is limited to special situations while in mainstream semiconductor device manufacturing, particle removal processes rely mostly on liquid-phase methods.

In contrast to particles, volatilization of residual organic contaminants physiosorbed at the wafer surface can be accomplished in a variety of ways by exposing the surface to the oxidizing agents such as ozone $O_3$ generated by UV illumination, breaking oxygen molecules and producing atomic oxygens ($O_3 \rightarrow O_2 + O$) featuring superior oxidation strength (Hoff and Ruzyllo, 1988). Also, processes carried out in oxidizing ambient at elevated temperatures, such as thermal oxidation of silicon for instance, were shown to be effective in organic contamination removal (Hossain *et al.*, 1990). As will be indicated in the later discussion, the ease with which organics can be removed using gas-phase treatments applies primarily to the situations concerned with relatively freshly physiosorbed compounds.

Regarding metallic surface contaminant removal, when using liquid-phase chemistries, metallic species are converted into water-soluble compounds and then rinsed away from the surface. In the case of gas-phase cleans, they need to be converted into volatile compounds or lifted off the surface by slight etching of silicon underneath (Daffron *et al.*, 1994). Both can be accomplished, for instance, through the reactions with UV-excited chlorine at the wafer temperature preferentially not exceeding 200°C (Fig. 6.10) (Chang *et al.*, 2004). However, experimental evidence suggests that reducing surface metallic contaminants using a UV/$Cl_2$ process to the required level ($<10^{10}$ atoms/$cm^2$) is not possible without nano-roughening of the Si surface and incurring material losses in the process.

This is because Cl-based chemistry cannot selectively interact with surface metallic contaminants without reacting with silicon at the same time (Ruzyllo *et al.*, 1998). As shown in Fig. 6.10, the temperature of the silicon wafer exposed to UV/$Cl_2$ at the pressure of 10 torr should not exceed 100°C to prevent excessive etching of silicon.

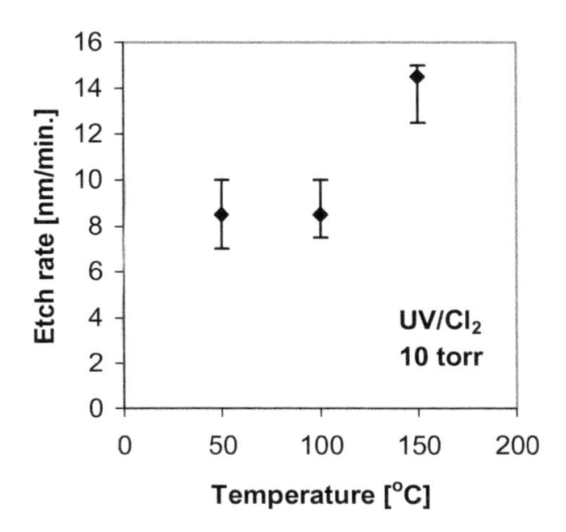


**Fig. 6.10.** Etch rate of silicon exposed to UV/$Cl_2$ as a function of temperature.

Moreover, the different vapor pressures of volatile compounds of various metals imply that not all metals can be equally effectively volatilized at low temperatures. The conclusion to be drawn from the above is that gas-phase chemistries are not effective enough in removing metallic contaminants from the Si surface across the desired range of metals without disturbing the cleaned surface.

As regards the removal of ultra-thin film of native silicon oxide $SiO_x$, grown on the silicon surface during wafer treatments, its removal prior to subsequent critical additive steps such as epitaxial growth or contact metallization is an inherent part of the surface processing procedures (Chin *et al.*, 1997). Considering its spontaneous, uncontrolled growth and the deleterious role it plays (see Fig. 3.9), the removal of such oxide from the surface using methods listed in Table 6.2 is a key part of the silicon surface treatments applied in the device manufacturing sequence (Roman *et al.*, 2009).

Some gas-phase native oxide removal methods are based on oxide sublimation in high vacuum and high temperature, as well as thermally stimulated reduction in $H_2$. Due to the high temperature needed to promote native/chemical oxide volatilization, both techniques can be employed only in selected surface processing sequences.

Other than the above, the options in terms of native/chemical oxide gas-phase etching that remain at our disposal fall into categories defined based on the chemistries used. It involves processes which continue to depend on the mechanism involving $H_2$ reduction of an oxide except that by using plasma rather than temperature to generate atomic hydrogen (Frystak and Ruzyllo, 1992), they can be implemented at a temperature as low as 300°C. The hydrogen plasma-based processes are well established both in terms of applications, including semiconductors other than silicon, as well as in terms of the commercial tool base.

Furthermore, the gas-phase processes (strictly speaking vapor-phase processes in this case) employed to etch native oxides use fluorine from anhydrous HF (AHF) to carry out oxide etching. Originally, water vapor needed to support the HF etch reaction in the vapor phase was directly mixed with vaporized AHF (Ruzyllo, 1993). The problem is that the AHF:$H_2O$ process leaves on the etched surface solid residues requiring follow-up water rinse to be removed. In the modified version of the process, vapors of alcoholic solvents such as methanol ($CH_3OH$, MeOH in short) or ethanol ($CH_3CH_2OH$) are used as a source of water needed to initiate etch reaction (Torek, Ruzyllo *et al.*, 1995). The AHF/alcoholic solvent processes suppress the formation of solid residues on the etched surface and improve etch selectivity as compared to the AHF/$H_2O$ vapor process. As such, they are successfully used in a range of oxide removal applications, including MEMS release processes (Erdamar *et al.*, 2008). In silicon surface processing applications, the AHF/MeOH vapor process is used to remove residual ultra-thin oxide from the surface prior to thin-film deposition steps.

The dependence of the AHF/MeOH etch rate on process parameters is illustrated in Fig. 6.11 which shows a correlation between the silicon oxide etch rate and pressure under which AHF/MeOH etching is carried out (Chang *et al.*, 2004). As indicated in the figure, the process is a strong function of pressure which in the case of removal of oxide films in the single nanometer thickness range needs to be lower than 200 torr to assure adequate process control.

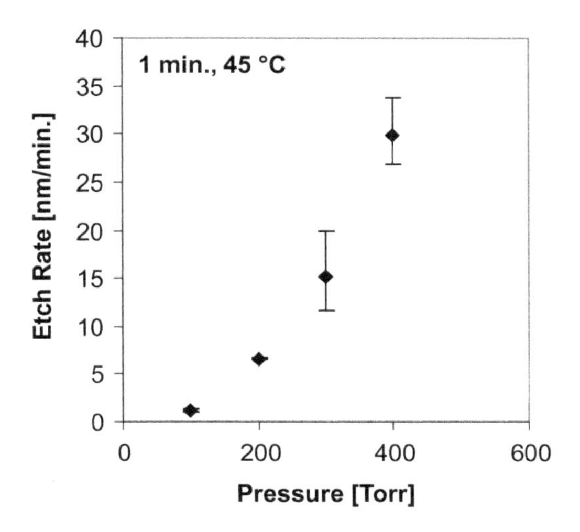


**Fig. 6.11.** The etch rate as a function of pressure during AHF/MeOH etching of thermal $SiO_2$.

In addition to the process pressure, wafer temperature has a strong effect on the AHF/MeOH etch rate. The etch rate decreases with wafer temperature regardless of pressure and at the wafer temperature of about 100°C and above, etching of the oxide does not take place (Ruzyllo, Torek *et al.*, 1993).

Yet another approach to gas-phase native oxide etching prior to contact metallization and epitaxial deposition involves remote plasma-generated $NF_3/NH_3$ chemistry. In this case, solid products of the etch reaction are removed from the surface by sublimation when the etching wafer temperature is increased above 100°C (Luther *et al.*, 1993).

Alternatively, the etching of silicon native oxide can be carried out in a way that prevents any damage to the exposed silicon surface using ultra-slow multicharged $Ar^{q+}$ ions (Le Roux *et al.*, 2004).

Regardless of the native/chemical spontaneously grown oxide etching method used, the process should be carried out *in situ* immediately prior to the subsequent deposition step whether it would be an epi layer, or metal, or high-*k* dielectric atomic layer deposition (ALD) process.

Overall, the gas-phase cleaning processes, regardless of what is their goal, should not be considered as an isolated step, but as a part of the surface processing procedure which takes advantage, as needed, of both liquid- and gas-phase chemistries (Ruzyllo, 2014). The latter benefits the process most when applied *in situ* as the surface conditioning step immediately prior to the subsequent deposition step carried out in the cluster tool.

## 6.4 Conditioning of Silicon Surfaces

The term "surface conditioning" refers to an operation or sequence of operations performed on semiconductor wafers to establish the desired chemical composition of its surface or to restore such conditions (Ruzyllo *et al.*, 1998; Shanmugasundaram *et al.*, 2005). Some elements of the broadly understood surface conditioning technology were considered earlier in this volume.

The point being made here is that surface cleaning may not be a final step in surface processing before the next process in the sequence which may require adequately processed substrate surface. In such instances, it is a common procedure to subject the surface to an additional surface conditioning process.

An example of a procedure falling into this category is a treatment-restoring condition of the clean surface after wafer handling in the clean-room air or exposure to an ambient during wafer storage or shipping. Among gas-phase methods allowing restoration of the condition of the surface subjected to handling and storage in the ambient air, the method of rapid optical surface treatment (ROST) (Kamieniecki, 2001; Tsai *et al.*, 2003; Danel *et al.*, 2003) was shown to be effective and compatible with semiconductor device manufacturing infrastructure.

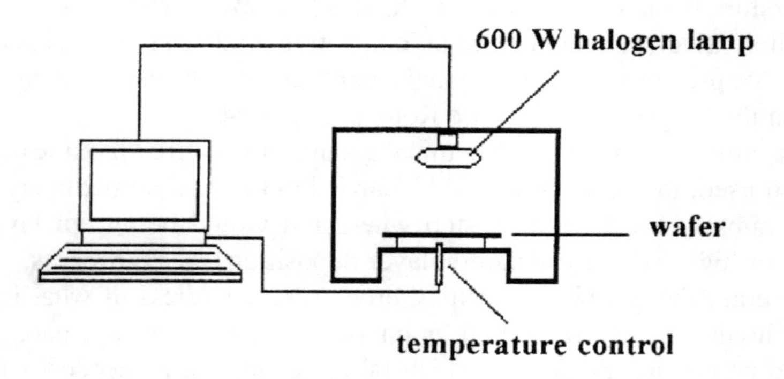


**Fig. 6.12.** Schematic representation of the ROST reactor.

As shown in the schematic representation of the ROST apparatus in Fig. 6.12, it uses a halogen lamp to generate white light which interacts with moisture and organics on the surface and at the same time increases the temperature of the wafer to the desired level not exceeding 300°C. The process, also referred to as lamp cleaning, is carried out in ambient air, and thus, the ROST reactor does not need to handle specialty gases and vacuum instrumentation. As such, it can be readily incorporated into the manufacturing line at any stage of the device fabrication process. Continuing the discussion of surface conditioning processes, it needs to be stressed that in the semiconductor device manufacturing environment what is considered a clean silicon surface may be a surface free from particles and metallic contaminants but covered with residual moisture and organics resulting from IPA wafer drying as well as exposure to the ambient air. During short-term exposure to ambient air, species on the surface remain physically adsorbed with no chemical interactions with the surface. During prolonged exposure of the surface to the ambient air, however, whether it is in a cleanroom, or in storage boxes, or in shipping containers, originally physiosorbed species gradually become chemically adsorbed as a result of moisture-catalyzed chemical reactions at the surface. The results and mechanism of this transition were considered in Chapter 3 and illustrated in Fig. 3.8 by the process of gradual changes in the chemical composition of silicon surface referred to as surface aging. From the semiconductor surface processing perspective, an additional conventional wet cleaning process is needed to take the chemically bonded contaminants of the surface and restore its chemical makeup to the desired condition.

The difference between the characteristics of silicon surface exposed to ambient air for short and long periods of time, and thus featuring mostly physiosorbed and chemisorbed organic contaminants, respectively, is reflected in the changes in the value of contact angle on the silicon surface subjected to the rapid optical surface treatment at the wafer temperature of 300°C.

The process can remove species adsorbed on the surface by means of thermal and optical stimulation, but only those adhering to the surface by weak forces of physisorption with no permanent chemical bonds between adsorbed species and the surface atoms formed during extended storage in ambient air. The process involving ROST is less effective when the surface is stored in the ambient air or shipping containers for a prolonged period of time, for instance, 10 months, as compared with the wafer fresh out of the shipping box (Fig. 6.13).

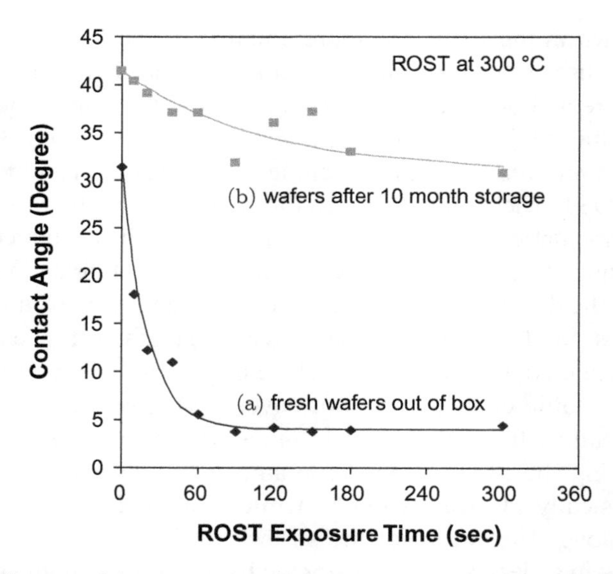


**Fig. 6.13.** Changes of the contact (wetting) angle on the Si surfaces of the wafers fresh out of the box and wafers stored in the cleanroom air for 10 months as a function of ROST exposure time at a wafer temperature of 300°C.

Contact angle results shown in Fig. 6.13 indicate that in the case of an Si wafer fresh out of the shipping box, its surface features a contact angle of 5° as a result of about 60-second-long ROST exposure. It means that the short exposure time was enough to remove physiosorbed organic contaminants and moisture from the surface and establish its hydrophilic characteristics. On the other hand, however, in the case of wafers stored in the cleanroom air for 10 months, even a 5-minute-long ROST is not enough to volatilize storage-related residues off the surface and instead a conventional wet cleaning may be needed to restore surface conditions.

Considering further ROST effectiveness in silicon surface conditioning processes, Fig. 6.14 illustrates the variations in the value of surface charge reflecting removal from the Si surface of the residues contributing negative charges to the surface (charge $Q_2$ in Fig. 4.8) and by doing this decreasing its impact on the total surface charge and causing positive charge associated with the structural features of silicon surface (charge $Q_1$ in Fig. 4.8) to dominate surface electrostatics (Tsai *et al.*, 2003).

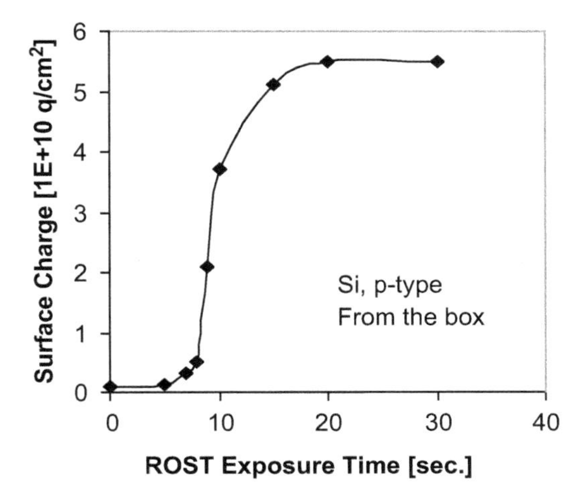


**Fig. 6.14.** Changes in the density of surface charge on *p*-type silicon wafer removed from the shipping box as a function of time of ROST exposure.

To reiterate points made earlier, the process of the semiconductor surface changing its chemical makeup with time of exposure to ambient air represents a lack of reproducibility of surface condition (Ruzyllo *et al.*, 1987, 1990). The solution to the problem of surface deposits accumulated during wafer shipment and storage assuring reproducibility of surface condition is a conventional surface cleaning carried out using liquid chemicals and deionized water. For this reason, surface cleaning of the shipped wafers is the very first step in a typical semiconductor device manufacturing sequence.

Several other examples of surface conditioning procedures applied in silicon device fabrication sequences can be given. In each case, the chemistry of the surface subjected to the conditioning process is determined by the needs of the subsequent operation. For instance, requirements regarding the chemical makeup of the surface prior to gate oxide formation and prior to contact deposition are different. Incidentally, the additive deposition steps are not the only ones that may require surface pre-treatment, considered here in terms of surface conditioning. For instance, prior to oxide etching using anhydrous HF/methanol vapor, the oxide surface is subjected to moisture-removing thermal treatment, such as ROST to ensure reproducible initiation of the etch reaction.

## 6.5 Control of Process-Induced Alterations of Semiconductor Surfaces

Other than contaminants requiring cleaning of semiconductor surfaces and interactions with process ambientes calling for surface conditioning, there is also a potential impact of the processes performed on the wafer during device manufacturing that may adversely affect its surface and near-surface region. Some processes employed in semiconductor device manufacturing, such as dry etching or implantation, may cause physical damage to the crystallographic structure of the surface of the processed wafer and alter the composition of the near-surface region. The special impact from this point of view has dry etching processes, including reactive ion etching (RIE) or magnetron enhanced RIE (MERIE) (Gu *et al.*, 1994), during which etched surfaces are exposed to the damaging interactions with energy-carrying ions (see comments in Section 3.5 of Chapter 3). Despite this, due to its anisotropy, RIE is a commonly applied etching mode in advanced semiconductor device manufacturing including processing of bare surfaces prior to the follow-up thin film deposition steps.

The downside of reactive ion etching is that it leaves the etched surface physically damaged and chemically altered which in most cases, depending on the next step in the device manufacturing sequence, require additional processing to minimize the adverse effect of these alterations (Moghadam and Mu, 2002). Figure 6.15 schematically illustrates the condition of the silicon surface following reactive ion etching. As seen in this figure, the RIE-related alterations of the surface condition include coverage of the surface with the layer of polymeric etch residues, structural damage caused by the acceleration of etching ions toward the surface, and contamination of the near-surface region originating from the etching gases and ionized gases interactions with the parts of the etching chamber (Hwang *et al.*, 1994). Also, post-RIE silicon features deep penetration by hydrogen from the etching gases which, as a fast diffusant in silicon is readily permeating its structure (Fonash, 1990).

In the process of restoration of the surface condition following RIE, polymeric etch residues remaining on the Si surface following RIE can be removed using various methods (Hwang *et al.*, 1996), including some considered earlier in this chapter. However, the elimination of surface damage concerned with the disruption of the crystallographic structure requires more involving additional treatments such as sacrificial oxidation

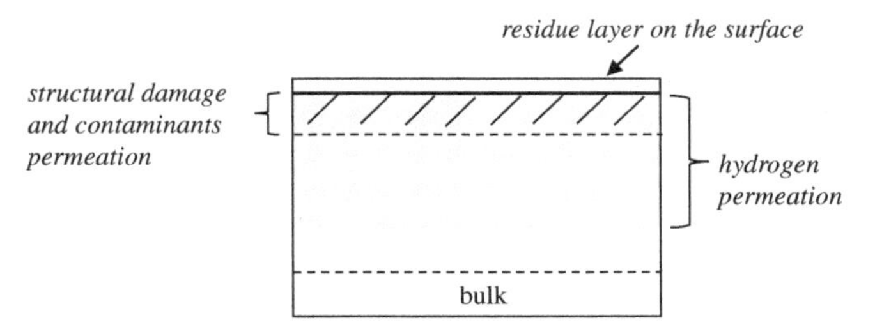


**Fig. 6.15.** Alterations of silicon surface and near-surface region introduced by RIE.

(see Fig. 3.5) or slight etching of silicon in the gas-phase using UV-stimulated chlorine etching of silicon, or UV/$Cl_2$ discussed in Section 6.1 (Hwang *et al.*, 1994). While in general, proven effective in post-RIE damage removal, the use of the former in post-RIE surface treatments needs to be considered keeping in mind that the elevated temperature associated with thermal oxidation may affect the chemical and physical characteristics of the near-surface region of silicon wafer.

With a trend in semiconductor device engineering toward 3D structure involving non-planar transistor configuration in advanced integrated circuits as well as in power transistors, both of which were discussed earlier in this *Guide*, additional challenges concerning etching technology and process-induced surface alterations come into play. As an example, selected aspects of surface characteristics in the case of power MOSFETs built into a U-shaped trench dry etched in silicon (UMOSFET) are considered next.

As demonstrated in Fig. 2.16, UMOSFET is a transistor in which the gate oxide and the channel run around the trench etched into a semiconductor substrate and filled with a gate contact material. Challenges associated with the processing of the MOS gate structure into a U-shaped trench can be seen in terms of those inherent to the shape of the MOS gate in this case as well as in terms of surface processing procedures carried out in the preparation of silicon surface in the trench prior to the growth of the gate oxide. To better understand the former, Fig. 6.16 shows a scanning electron microscope image of a 0.4 $\mu$m wide and 2 $\mu$m deep trench (Fig. 6.16(a)) and a magnified view of its bottom part with gate oxide grown around its edge and a trench filled with poly-Si acting as a gate contact (Fig. 6.16(b)) (Grebs *et al.*, 2004).

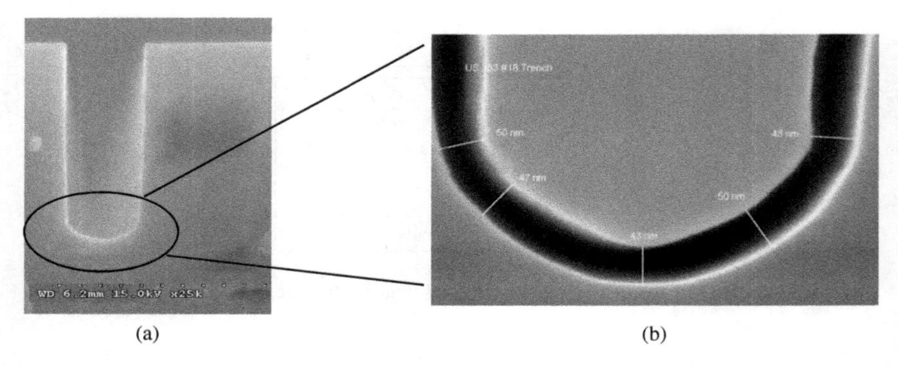


**Fig. 6.16.** (a) Scanning electron microscopy view of the U-shaped trench reactive ion etched into silicon and (b) part of it showing non-uniform thickness of the thermal oxide grown inside the trench.

As a cross-section of the trench in Fig. 6.16 demonstrates, different gate oxide thicknesses in the sections corresponding to the differences in the crystallographic orientation of the surface on which oxide was thermally grown can be distinguished. Experimentally observed resulting reduced oxide breakdown field in the trench as compared to the oxide grown under exact same conditions on the flat surfaces is due to the non-uniform oxide thickness seen in Fig. 6.16(b), the stress in the oxide at various points along the edge of the trench as well as different average oxide thickness in the trench and on the flat surface (Chang *et al.*, 2004).

In terms of the pre-gate oxidation treatments of the surface in the trench, measurements of the breakdown field of the oxides thermally grown in the trench on the surface prepared by means of sacrificial oxidation and by using conventional Si surface treatments have demonstrated superior oxide breakdown statistics in the case of trench capacitors formed on the surface first subjected to sacrificial oxidation (Fig. 6.17) (Grebs *et al.*, 2004).

The conclusion that can be drawn from the discussion in this section is that in contrast to the chemical condition of the surface, which can be controlled using methods discussed earlier in this chapter, physical damage to the surface and near-surface region caused by reactive ion etching in most cases cannot be removed without additional processes beyond surface cleaning. Also, the processing of physically altered surfaces in three-dimensional semiconductor structures such as narrow trenches may require different procedures than in the case of planar surfaces. Among

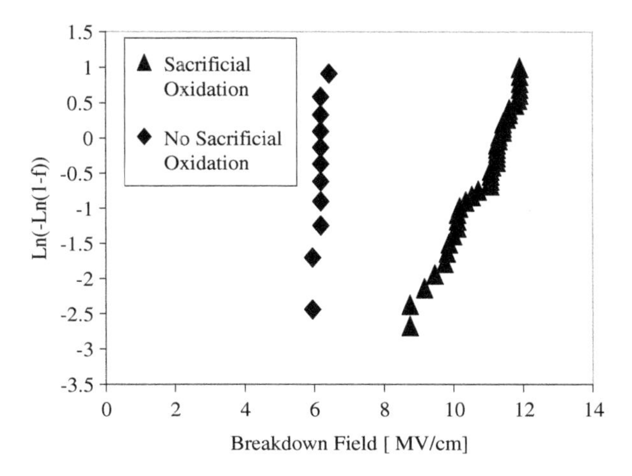


**Fig. 6.17.** Breakdown statistics of the silicon dioxide thermally grown in the trench with and without prior sacrificial oxidation process.

the reasons for these differences is the inability of liquid-phase chemistries to uniformly penetrate deep surface features due to the surface tension. In some cases, supercritical cleaning can be a solution, but in others, slight etching of silicon, for instance by means of gas-phase treatments such as UV/$Cl_2$ may be an option.

## 6.6 Surface Processing of Substrate Materials Other than Silicon

It is well established that due to its high crystal quality, adequate electronic properties, abundance, excellent oxidation characteristics, manufacturability, relatively low cost, and availability in the form of large wafers, silicon is a dominant semiconductor material used in the manufacture of electronic devices. However, growing needs for improved performance in specific electronic applications in terms of power handling, temperature resistance, and speed of operation, as well as in photonic applications in terms of emission of blue light, UV detection, and efficiency of photovoltaic effect, call for the mature surface processing technology in the range of semiconductor materials other than silicon. Consequently, surface processing is also an issue of interest in the commercial applications of semiconductor materials other than silicon. This is also because, in light of improvements in the quality of various single-crystal semiconductor substrates, there is a growing recognition of the effect of surface characteristics on device manufacturing yield and performance.

In general, technology of surface processing, including cleaning, of materials other than silicon attempts to draw from the pool of silicon cleaning chemistries using high-performance surface processing infrastructure and characterization methods developed for silicon. While there are no barriers regarding the latter, the selection of cleaning chemistries for various other semiconductor materials is not a straightforward matter. This is because due to the inherent differences in the chemical composition, various semiconductor materials may not adequately respond to the surface cleaning and conditioning chemistries successfully used in silicon processing.

In this section, examples of surface processing-related issues in the case of single-crystal germanium, silicon carbide SiC, as well as selected III–V semiconductor compounds are briefly reviewed. In addition, issues related to the surface processing of sapphire used as a substrate for silicon in silicon-on-sapphire (SOS) integrated circuits as well as for epitaxially deposited GaN are addressed.

**Germanium surface processing.** As it features higher than in the case of silicon mobility of electrons and holes, germanium is an alternative to silicon elemental semiconductors in applications in which the frequency at which devices operate is of importance. On the other hand, inferior properties, including solubility in water of its native oxide of $GeO_2$, restrict the use of germanium in the MOS device technology with the exception of applications where Ge alloyed with Si (silicon germanium, SiGe) is used to induce strain increasing electrons' mobility in the MOSFET's channel.

As far as standard cleaning operations such as particle and metallic contaminants removal and residual native oxide etching are concerned, the response of the Ge surface to the Si cleaning chemistries varies depending on the application. In the case of particle removal, for instance, the germanium surface acts in the same way as the silicon surface. Similarly, organic contaminants can be removed from the Ge surface using the same techniques as in the case of silicon. On the other hand, metallic contaminant removal was shown to be driven in the case of Ge by somewhat different mechanisms than in the case of Si substrates. Most notably, unlike in the case of Si, common in semiconductor technology metallic contaminants can be removed from the Ge surface using HF:$H_2O$ solution.

Other than the above, similar to some other semiconductor materials, germanium surface features sensitivity to prolonged exposure to hot water. The effect is demonstrated in Fig. 4.4 which shows an increase in the roughness of the germanium surface from RMS 0.318 nm to 1.04 nm as a result of prolonged immersion in boiling water.

**Silicon carbide surface processing.** Silicon carbide (SiC) is a wide bandgap semiconductor material of choice in power device applications. It crystallizes in hexagonal unit cells of various polytypes among which 4H and 6H polytypes are commonly used in SiC device manufacturing. In contrast to elemental semiconductors such as silicon and germanium, silicon carbide represents a class of synthetic binary semiconductor compounds in which comprising elements typically feature different chemical characteristics. Hence, the response of compound semiconductors to surface-affecting treatments may not be entirely isotropic. In the case of SiC, for instance, during thermal oxidation, oxidized Si atoms in SiC form a solid $SiO_2$, while oxidized carbon forms gaseous compounds CO and $CO_2$, effectively removing carbon atoms from the oxidized surface of silicon carbide (see the discussion in Section 2.4). The result of this treatment was schematically illustrated in Fig. 2.9 while atomic force microscopy (AFM) images in Fig. 6.18 show the extent of surface roughness resulting from thermal oxidation of SiC followed by oxide etching (Chang *et al.*, 2005). Similar effect of thermal treatments can be observed in other compound semiconductor materials as well (Nishimura, 1991).

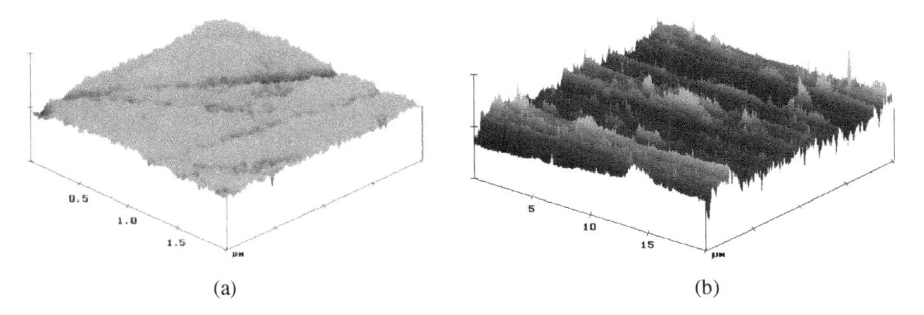


**Fig. 6.18.** AFM image of SiC surface after (a) thermal oxidation and (b) oxide stripping by etching in HF:$H_2O$ solution.

In spite of the binary nature of SiC compound, due to the fact that SiC is a chemical derivative of Si, cleaning chemistries used in Si processing are adopted, rather arbitrarily, to process SiC surfaces. It turns out that such an automatic transfer of cleaning technology from Si to SiC may not necessarily produce the desired results. For instance, variations of SiC surface roughness exposed to various cleaning chemistries used in Si technology were observed (Chang *et al.*, 2005).

Considering the hardness of silicon carbide and resulting difficulties in the removal of surface damage incurred by polishing during SiC wafer fabrication (see the discussion in Chapter 4), it is a common procedure to cover the physically damaged surface of as-processed SiC wafers (see Fig. 4.5) with the epitaxial layer of silicon carbide thick enough to minimize adverse effects by the physical damage of the surface of the substrate wafer.

**III–V compound surface processing.** Considerations suggesting the need for surface cleaning and conditioning specific to any given chemical composition of compound semiconductor, also apply to III–V materials. In the case of crystalline III–V compound semiconductors which are used in both electronic and photonic device applications, there are some, for instance gallium arsenide (GaAs), in which surface processing-related challenges are well recognized and controlled (Baca and Ashby, 2005). In other cases, ongoing research on the way compounds featuring diverse chemical compositions respond to various surface treatments is responsible for the improved understanding of surface processing related issues.

As an example, Fig. 6.19 shows the changes in the minority carrier lifetime in the near-surface region of single-crystal gallium antimonide (GaSb) subjected to various treatments preparing surface to atomic layer deposition of 5 nm thick layer of $Al_2O_3$ (Hwang *et al.*, 2011).

Surface treatments evaluated included decrease in acetone and isopropanol alcohol for 5 min (Process 1), followed by etching in HCl(1):$H_2O$(2) solution for 10 min, and then rinsing in isopropanol alcohol (Process 2). This last treatment resulted in damage to the GaSb surface expressed by the decreased minority carrier lifetime in the near-surface region of the wafer. However, when followed by the low-temperature annealing (Process 2 + anneal), no significant reduction of the minority carrier lifetime was observed.

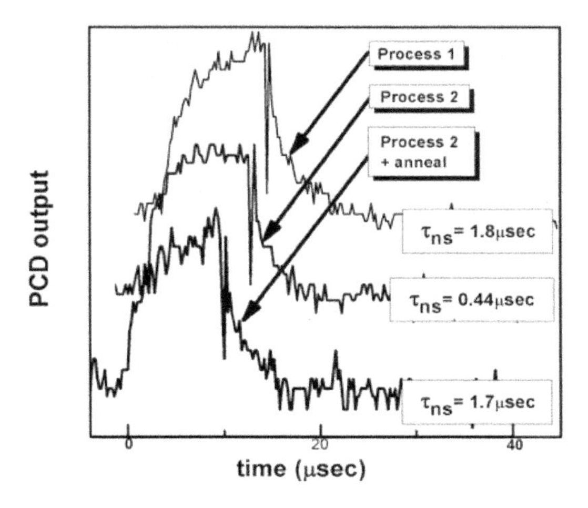


**Fig. 6.19.** Minority carrier lifetime in gallium antimonide, GaSb, subjected to three different surface treatments defined above in the text, applied prior to atomic layer deposition of 5 nm thick $Al_2O_3$.

Among other III–V semiconductor compounds, in the case of gallium nitride (GaN), issues related to surface processing technology attract significant attention (Chu, 2021). It is not only due to the importance of this wide and direct bandgap semiconductor material in power devices and LED technology but also because unlike other semiconductor materials considered in this discussion, in which the crystallographic structure is formed based on elemental cubic cells, gallium nitride crystalizes in the structure based on the hexagonal cells (Fig. 2.6). This characteristic of GaN plays a role in the discussion of surface processing in the case of this material as the crystal planes representing the surface in the complex hexagonal structure, upon which devices are formed, predetermine responses to various surface treatments.

For additional current information regarding the surface processing of semiconductor materials other than silicon, readers are referred to proceedings from the symposia related to the processing of semiconductor surfaces (Symposia, 1989–2025, 1992–2025).

**Sapphire surface processing.** As was alluded to earlier in this *Guide*, there are high-performance, advanced semiconductor devices formed on single-crystal insulating substrates. The prime example here is sapphire (single-crystal alumina $Al_2O_3$) which distinguishes itself with superior resistance to temperature and aggressive chemistries as well as transparency to light within a broad range of wavelengths. Due to these advantageous characteristics, sapphire is used in silicon-on-sapphire (SOS) wafers representing variations of silicon-on-insulator (SOI) technology used to manufacture devices and circuits designed for operation at high frequency, which are built into silicon active layer deposited on the sapphire substrate (Arora *et al.*, 2016).

Furthermore, sapphire is commonly used as a substrate for heteroepitaxial deposition of GaN to form wafers compatible with the needs of large-scale commercial manufacturing of GaN-based devices. In both applications, prior to the deposition of semiconductor material, the surface of the sapphire substrate needs to be subjected to surface processing operations.

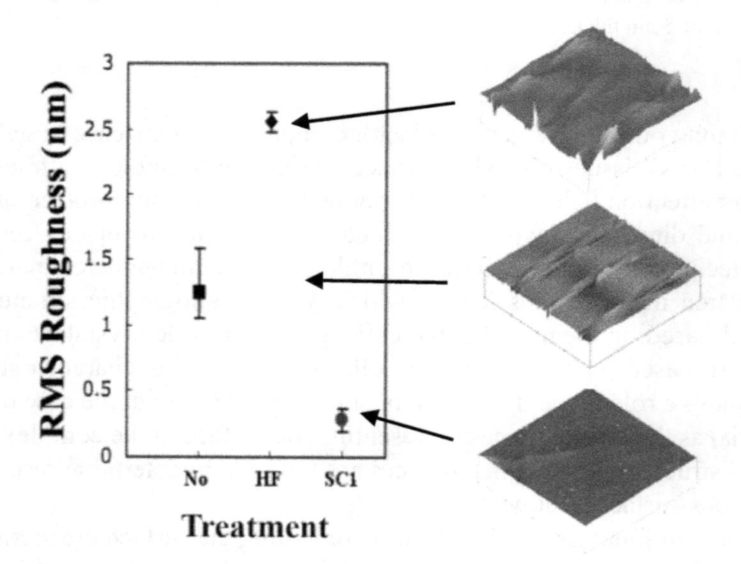


**Fig. 6.20.** Effect of HF and SC-1 solutions used in silicon processing on the roughness of sapphire surface compared to untreated surface.

The results of exposure of the sapphire surface to standard wet cleaning solutions used in silicon technology discussed earlier in this chapter are shown in Fig. 6.20 in terms of the effect on the roughness of the sapphire surface (Kirby *et al.*, 2007). As seen in this figure, treatment of as-received sapphire wafer in HF:$H_2O$ roughens its surface significantly. On the other hand, treatment in SC-1 solution renders the sapphire surface smoother. What it is telling us is that the surface of sapphire responds differently from silicon to the treatments using solutions commonly used in silicon cleaning, and therefore, sapphire-specific cleaning solutions need to be developed.

As it can be concluded from an overview of the surface processing of materials used in the manufacture of semiconductor devices other than silicon, there are instances where silicon cleaning solutions work well with other semiconductor materials, and there are others when the opposite is observed. In general, however, surface cleaning and conditioning methods need to be selected keeping specific characteristics of any given semiconductor material in mind.

# Chapter 7

# Semiconductor Surface Characterization Methods

## 7.1 Introduction

The goal of this chapter is to discuss material characterization methods used for the purpose of monitoring semiconductor surface conditions in broadly understood semiconductor device technology. In agreement with the scope of this book, the emphasis in this discussion is on the methods selected based on compatibility with applied semiconductor research and general industrial practice. In combination, these methods provide information regarding the condition of the surface of the processed wafers sufficient to monitor the outcome of the processes employed in semiconductor device manufacturing. Readers interested in more general, in-depth coverage of surface characterization techniques are referred to other sources covering surface characterization methods and their scientific background such as, for instance, the works by Dabrowski and Mussig (2000), Hattori (1998), Irene (2008), McGuire (1989), Mönch (2001), and Schroder (2015).

In the review of methods of surface characterization used in semiconductor device engineering, the techniques of interest are considered in the classes distinguished based on the principles of their operation. First, methods commonly referred to as analytical techniques involve X-ray irradiation of the surface, beam of electrons, or accelerated ions impinging on the surface. Then, ellipsometry which belongs to the class of optical methods is considered as a technique allowing extraction of the information regarding specific characteristics of semiconductor surfaces.

In the remaining sections of this chapter, the principles of atomic force spectroscopy (AFM) and the determination of variability of characteristics of the as-processed surfaces based on the wetting (contact) angle measurements are considered. The final section is focused on electrical methods considered from the point of view of their effectiveness, specifically in the characterization of the bare semiconductor surface and near-surface region as well as their usefulness in the monitoring of surface processes in semiconductor device manufacturing.

Throughout the discussion in this chapter, silicon surface characterization methodologies are used as an example. As the most often characterized, and thus the best understood, silicon surface provides an adequate platform upon which considerations of the merits and challenges of semiconductor surface characterization methods can be based.

## 7.2 Surface Characterization Techniques in Semiconductor Device Technology

In this section, semiconductor surface characterization methods used in applications concerned with semiconductor device technology are considered. The selection of techniques discussed here is based on their perceived usefulness in the characterization of semiconductor surfaces in various practical applications, as well as on the author's research experiences in this area. Classes of methods commonly used in semiconductor surface characterization and selected for the discussion in this overview are introduced in Table 7.1. Methods listed in this table are first considered in this section in general terms and then discussed in more detail later in this chapter. These methods, in various combinations, are believed to provide information regarding the condition of the semiconductor surface sufficient to assess the performance of fabrication processes in terms of their impact on surface characteristics. Of interest in this overview are techniques listed in Table 7.1 featuring depth resolution appropriate for the characterization of the surface and near-surface region of the semiconductor wafer. Also, the focus of the follow-up discussion is on the methods able to distinguish surface and near-surface effects from the semiconductor bulk-controlled phenomena.

**Table 7.1.** Surface characterization methods selected for consideration in this discussion.

| Analytical Methods | Optical Methods | Electrical Methods | Supporting Methods |
|---|---|---|---|
| *XPS | *Ellipsometry | *Non-contact | *Atomic Force Microscopy (AFM) |
| *TXRF | | *Temporary contact | * Wetting (contact) angle measurements |
| *AES | | | |
| *TOF-SIMS | | | |

**Analytical methods.** The term "analytical" is used here in reference to surface analysis techniques providing qualitative and quantitative information regarding the chemical composition of the surface. Regardless of the method used, to obtain useful information concerning the characteristics of the solid surface, the energy needed to stimulate the desired reactions must be delivered to such a surface. The energy, if delivered to the surface in the right way, initiates interactions with exposed material which, when detected and analyzed, provide useful in practice information regarding the condition of the surface and near-surface region.

To obtain surface and near-surface specific information, incoming energy needs to be absorbed within the shallow region adjacent to the surface. The number defining the energy absorption depth needs to be considered in the context of the method used, the semiconductor material characterized, and the expected extent of surface region alterations. For the purpose of this discussion, it is assumed that it should be within some 100 nm from the surface.

Interactions associated with the use of analytic techniques in semiconductor surface characterization are presented in a simplified fashion in Fig. 7.1. The energy delivered to the surface, referred to as activation energy, is absorbed at the surface causing the release of a product carrying information regarding various features of the surface and near-surface region. This can be, for instance, information concerning the chemical composition of the region in which the incoming energy was absorbed. To accomplish these tasks, the product of the surface reaction needs to be detected using dedicated detectors and analyzed using a signal processing system.

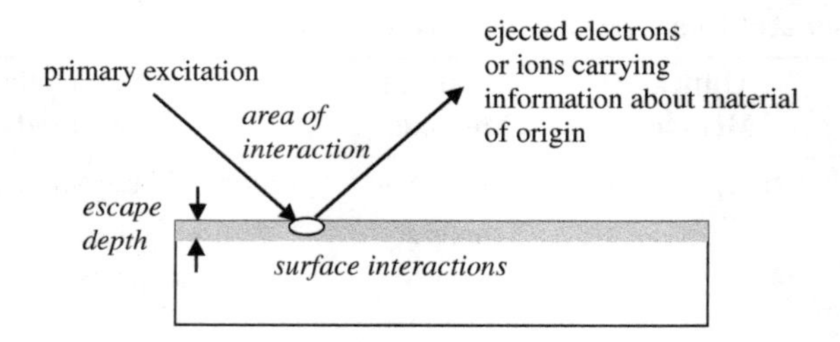


**Fig. 7.1.** Surface and near-surface interactions underlying working of surface analysis methods.

The way energy is delivered to the surface to initiate the process illustrated in Fig. 7.1 is the basis upon which methods of surface analysis are distinguished and grouped into various classes. As indicated earlier, the excitation energy can be in the form of X-ray radiation, kinetic energy of accelerated electrons focused into beam, or a beam of chemically inert ions, typically argon ions, impinging on the surface. Whether the results carrying information about the condition of the surface are detected in the form of the spectra of emitted electromagnetic energy, energy of emitted electrons, or mass of the sputtered ions, the approach illustrated in Fig. 7.1 is a foundation of the broad range of analytical methods applied in semiconductor surface analysis.

What needs to be noted is that to prevent undesired interactions of excitation energy-carrying species, as well as emitted products of the surface interactions with gas molecules in the ambient, all these methods need to be carried out in the vacuum environment. This is a consideration determining where and how analytical techniques are being used in surface characterization in semiconductor device technology.

Additional introductory comments on earlier mentioned methods that belong to this class and are distinguished based on the nature excitation source including X-rays, electrons, and ions are as follows.

Depending on how it is applied, the use of X-ray irradiation of crystalline solids can provide useful information regarding the investigated material based on the emission of photoelectrons, on the effect of fluorescence, or on the analysis of diffraction of the X-rays irradiating studied material. In the case of surface characterization, X-ray excitation needs to be limited to the region immediately adjacent to the surface of the irradiated solid so that signal-carrying information regarding its condition originates from the surface region only. In the case of X-ray-stimulated emission of photoelectrons, the determination of their origin is based on their energy (Fig. 7.2(a)).

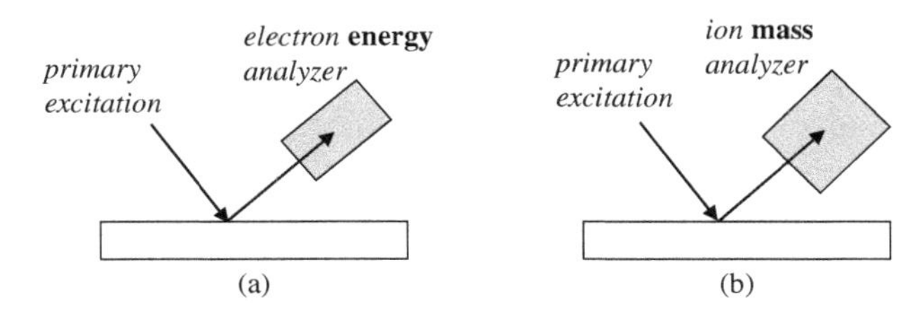


**Fig. 7.2.** (a) Origin of emitted electrons is based on the analysis of their energy and (b) in the case of sputtered ions, their origin is determined based on their mass.

The same applies to the techniques used in semiconductor surface characterization which are based on electron excitation. A finely focused electron beam impinging on the surface causes the emission of secondary electrons which carry information about the topographical features of the surface or the chemical origin of the atoms from which electrons were ejected. Similar to X-ray excitation, in this case the outcome of the process is also determined by analyzing the energy of the emitted electrons using energy analyzers integrated with the surface characterization setup in the way shown in Fig. 7.2(a).

In the case of ion beam excitation, the beam of heavy ions, typically argon $Ar^+$, is bombarding the surface, transferring the energy through collisions with atoms in the exposed surface region and causing ejection of ions from the investigated material. In contrast to methods based on X-rays or electron beams enforcing energy interactions within the studied material, in the case of ion excitation the origin of sputtered ions is determined based on their mass. Sputtered ions are captured by the mass spectrometer which determines the mass and then, based on the established standards, their origin (Fig. 7.2(b)).

As will be revealed in the discussion in Section 7.2, the analytical methods explained in general terms above are modified in semiconductor practice when the analysis needs to be limited to the surface and near-surface region. For instance, an angle at which X-rays arrive at the analyzed surface can be modified to ensure sufficiently shallow penetration depth. In the case of ion-beam-based processes, the type of the primary beam and its energy can be changed to ensure interactions with the analyzed surface are limited to the near-surface region only.

**Optical methods.** Another class of surface characterization methods involves optical methods which use as an activation energy radiation featuring wavelengths from the UV, visible, and infrared parts of the electromagnetic spectrum, depending on the needs (Saleh and Teich, 1991). Featuring lower energy than X-ray radiation, the optical methods of material characterization do not rely on the analysis of species ejected from the near-surface region, but instead involve the detection and analysis of optical effects involving light such as absorption, scattering, diffraction, interference, and reflection to identify features of the illuminated surface of semiconductor material.

In terms of the absorption of light by the illuminated material (Fig. 7.3(a)), a correlation of the absorption coefficient defining the depth of penetration of light in a specific semiconductor material and the wavelength of light illuminating its surface needs to be noted (Fig. 3.11). In general, methods using wavelength from the visible and infrared (IR) range do not provide surface-specific information. The infrared (IR) based methods such as IR spectroscopy, including Fourier transfer infrared (FTIR) spectroscopy, while being very useful and broadly used in material characterization, do not allow the separation of surface-related phenomena. For instance, as shown in Fig. 3.11, the absorption depth in silicon of the infrared light featuring 800 nm wavelength is about 12 $\mu$m, which is beyond the extent of the surface and near-surface region.

The effect of light scattering in turn (Fig. 7.3(b)) typically involving wavelength in the 500–1000 nm range is exploited in particle counting applications. As discussed earlier in this volume, particles adsorbed at the surface of the processed semiconductor wafer are responsible for the major defects-causing effects in semiconductor technology, and thus, need to be detected and counted. In particle counters, the light shed on the surface of the wafer is scattered on the particles present there, and the resulting light points are detected and counted by the instrument. The procedure is mostly dependent upon the optical phenomena involved and not on the condition of the particles' contaminated surface. Therefore, light diffraction-related particle counting procedures, discussed in the literature, for instance (Hattori, 1998; Kern, 1993), are not included in the consideration of optical methods of surface characterization in this *Guide*.

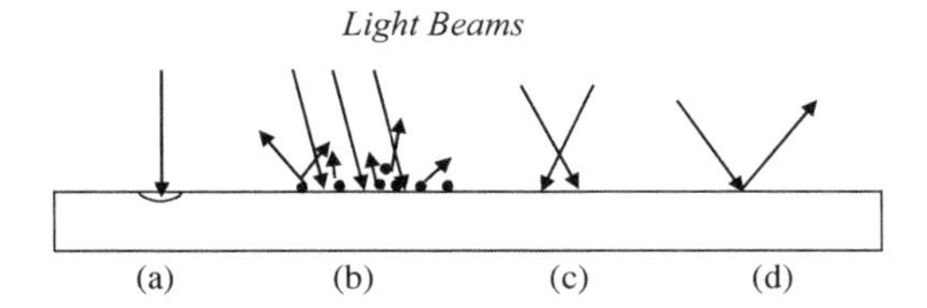


**Fig. 7.3.** Illustration of optical effects occurring during light beam interactions with solid surface: (a) absorption, (b) diffraction-related scattering, (c) interference, and (d) reflection.

Another optical effect used in semiconductor material characterization involves interference between superimposed beams of short-wavelength light (Fig. 7.3(c)). Laser interferometry, capable of detecting changes in the optical characteristics of irradiated material, is used for end-point detection in dry etching operations. While readily detecting changes in optical characteristics from material to material during dry etching, laser interferometry in its basic configuration is not used to detect fine changes in the characteristics of single-crystal semiconductor surfaces.

Keeping the above constraints in mind, the effect of light reflection (Fig. 7.3(d)) is considered as an appropriate foundation for the optical characterization of semiconductor surfaces. The method of ellipsometry which is based on the analysis of light reflection from the monitored surface is considered in Section 7.3 as a technique representing optical methods compatible with the characterization of the bare semiconductor surfaces, as understood in the discussion in this volume.

**Atomic Force Microscopy (AFM).** The AFM method is among the surface characterization techniques used in semiconductor technology defined in Table 7.1 as supporting methods. Included in the discussion in Section 7.4, the method of AFM is broadly used in semiconductor technology to evaluate surface roughness with atomic-scale precision (Sanders, 2019). Based on attractive and repulsive forces, AFM provides information regarding the physical condition of the surface with effectiveness unattainable using other methods.

**Wetting (contact) angle measurements.** As demonstrated earlier in Section 4.1, surface energy is directly related to the condition of solid surfaces including the condition of single-crystal semiconductor surfaces of interest in this book. Changes in the value of surface energy in turn affect the wettability of the surface, which is quantitatively represented by a wetting angle, also referred to as a contact angle (Fig. 4.2). Contact angle can be measured when a droplet of liquid, for instance, water, is placed on the characterized surface and because of the surface tension of the liquid and the attraction of the liquid to the solid surface, the droplet assumes a dome shape allowing measurement of the wetting angle using a setup shown later in this chapter.

The adequately measured value of the contact angle represents changes in the surface energy of semiconductor material subjected to various treatments. The contact angle is directly related to the chemical and physical characteristics of the solid surfaces and its measurements can be used to follow changes in the condition of semiconductor surfaces during the device manufacturing procedures.

The discussion in Section 7.5 considers the practical aspects of contact angle measurements used in semiconductor engineering. More detailed information regarding the physics of contact angle measurements is available in the literature (Decker and Garoff, 1999).

**Electrical methods.** The performance of electronic semiconductor devices such as transistors or diodes, as well as selected photonic devices such as light-emitting diodes is predetermined by the semiconductor material system sensitivity to electrical stimulation upon which operation of such devices is based (Ruzyllo and Roman, 1999). Compared to other methods of material characterization, a key advantage of electrical methods is that the results obtained are directly correlated to the performance of the final semiconductor device (Caymax *et al.*, 1998).

For instance, reduced by a defective material, electron mobility or minority carrier's lifetime measured by means of electrical techniques will be directly reflected in the inferior performance of the device manufactured using this material. At the other end, the impact of features of semiconductor surfaces determined, for instance, by means of analytical techniques considered earlier, in most cases requires additional experimentation and specialized knowledge in order to establish its effect on device performance. For these reasons, methods of electrical characterization serve unique purposes in semiconductor device engineering including applications in process monitoring (Ruzyllo and Drummond, 2016).

There are electrical characteristics of semiconductor surfaces, for instance density of surface states or surface recombination velocity, that can be measured using various methods. Therefore, a measured property may not be an adequate criterion to distinguish between measuring techniques employed. Instead, the type of contact needed to drive the electrical signal in and out of the measured sample and the configuration of the test structure can be used to distinguish between measurement methodologies and the way they are used. In terms of the former, electrical methods of surface characterization fall into the categories of techniques using permanent contact to the surface (Fig. 7.4(a)), temporary contact (Fig. 7.4(b)), and non-contact methods which perform measurements without physical contact to the surface (Fig. 7.4(c)).

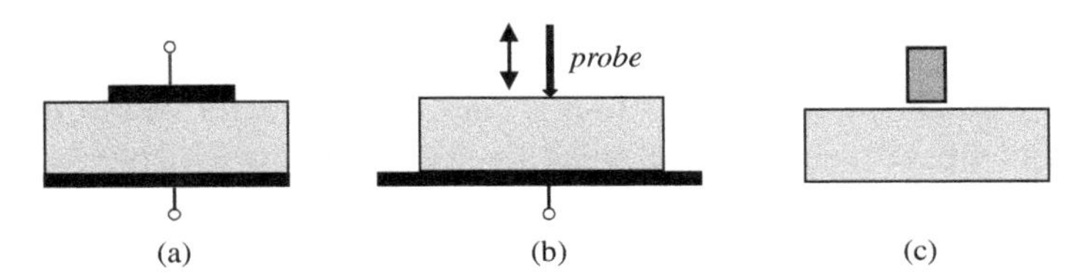


**Fig. 7.4.** Possible contact configurations in electrical characterization of semiconductor surfaces including (a) permanent contact, (b) temporary contact, and (c) non-contact methods.

Test structures based on permanent contact involve metal contact deposition by, for instance, vacuum evaporation. A temporary point contact to the measured surface can be formed using probes made of hard metal such as tungsten, or larger area temporary contact using soft metal probes. Alternatively, some electrical characteristics of semiconductor surfaces and near-surface regions can be obtained using non-contact methods in which the connection between the measuring probe and the surface is established by capacitive coupling, for instance.

With the discussion in this book focused on the monitoring of the condition of as-processed surfaces, or in other words surfaces which were not subjected to any additional treatments prior to the measurement, test structures involving permanent contact are not considered in the follow-up discussion. This includes MOS test structures requiring growth of the gate oxide and contact deposition, or Schottky diodes and ohmic contacts in which metal deposition is preceded by specialized surface treatments. Techniques of semiconductor surface and interface characterization involving test structures with permanent contacts are broadly considered in the literature (Irene, 2008; Ruzyllo and Roman, 1999; Schroder, 2015).

Based on the above introductory comments, a more detailed discussion of electrical methods of semiconductor surface characterization in Section 7.6 will be limited to the techniques using temporary contact and non-contact methods to derive information regarding surface characteristics affecting the performance of semiconductor devices. In agreement with the focus of this book, only methods used to characterize bare semiconductor surfaces in device engineering applications are considered.

## 7.3 Selected Analytical Methods Based on X-ray, Electron, and Ion Excitation

As discussed above, techniques in this category characterize semiconductor surfaces based on the principles of spectroscopy which rely on the determination of the spectra produced because of solids' interactions with very short-wavelength electromagnetic radiation such as X-rays or energy-carrying particles like electrons and ions. To ensure interactions of this nature are not disturbed, most of the surface characterization methods in this category need to be carried out in a vacuum environment which has an effect on the way and for what purpose they are applied in semiconductor device engineering.

Among X-ray-based analysis methods, X-ray diffraction (XRD) analysis which determines the intensities and scattering angles of the X-rays leaving irradiated material is not used specifically to analyze the features of semiconductor surface and near-surface region. Therefore, while it is important in the studies of the crystallographic structure of the solids in general, X-ray diffraction (XRD) analysis is not included in the follow-up discussion of surface characterization methods concerned specifically with X-ray analysis of semiconductor surfaces in which X-ray photoelectron spectroscopy (XPS) and total reflection X-ray fluorescence (TRXF) spectroscopy are highlighted.

Similar to X-ray diffraction considerations, apply to frequently used in material characterization methods based on electron diffraction. Due to their shallower penetration depth, electron diffraction techniques are the preferred methods applied, for instance, in the monitoring of the growth of the ultra-thin epitaxial films. The choice between low-energy electron diffraction (LEED) (Lander and Morrison, 1962), high-energy electron diffraction (HEED), or reflection high-energy electron diffraction (RHEED) depends on applications which are primarily related to the analysis of the crystallographic structure of semiconductor materials and do not include monitoring of the condition of semiconductor surfaces during device manufacturing procedures. As an example, the RHEED technique is often used in the monitoring of molecular beam epitaxy (MBE) discussed in Section 8.4.

Keeping all the above in mind, further discussion in this section is focused on the spectroscopic methods which are best compatible with the needs of semiconductor surface characterization in device manufacturing.

**X-ray Photoelectron Spectroscopy (XPS).** The XPS, also known as electron spectroscopy for chemical analysis (ESCA), is a technique routinely used in the characterization of semiconductor surfaces. This is because based on the principles of its operation, XPS differentiates between the characteristics of the bulk and the surface of the semiconductor substrate due to the fact that photoelectrons can be ejected from the solid only from the region immediately adjacent to the surface. As a result, XPS can provide information regarding the specific chemical composition of the studied surface. To ensure uninterrupted access of X-rays to the studied surface, as well as emitted photoelectrons to the electron energy analyzer, XPS analysis needs to be carried out under high-vacuum conditions.

Figure 7.5 illustrates the principles of X-ray photoelectron analysis which is based on deep-level electron emission driven by X-ray excitation. Primary excitation resulting from the X-ray irradiation of the studied surface involves X-ray photons that carry sufficient energy, typically in excess of 1 keV, to eject an electron from the atomic core levels. The core level energy spectra determined by an electron energy analyzer yield information about the identity of the atoms from which electrons are emitted.

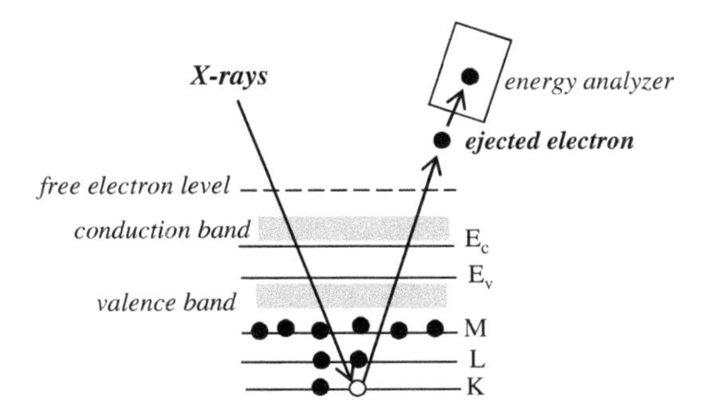


**Fig. 7.5.** Inter-level transitions upon which operation of X-ray photoelectron spectroscopy (XPS) is based.

The number of electrons recorded for a given transition is proportional to the number of atoms at the surface which allows XPS to provide both qualitative and quantitative information regarding atoms on the surface from which electrons are ejected. At the escape depth of X-rays generated electrons in the range of a few nanometers, the XPS information is derived from the atoms on the surface and in the near-surface region which makes XPS the broadly used surface analysis technique.

The main use of XPS in semiconductor surface characterization is to identify atomic species common in the semiconductor process environment, such as oxygen, carbon, or fluorine adsorbed on the semiconductor surface. Most of the other elements can be detected using XPS as well. A notable exception is hydrogen which with only one electron in its atomic structure does not lend itself to energy exchanges upon which the operation of XPS is based. On the other hand, molecular species such as hydrocarbons can be detected using XPS. This, in conjunction with XPS's ability to provide information regarding binding energies of surface species, adds significantly to the versatility of this surface characterization technique. As an example of XPS use in applications of interest in this discussion, Fig. 7.6 shows the XPS spectrum identifying silicon, carbon, and oxygen on the surface of an Si wafer stored in the ambient air and covered with spontaneously grown ultra-thin silicon oxide.

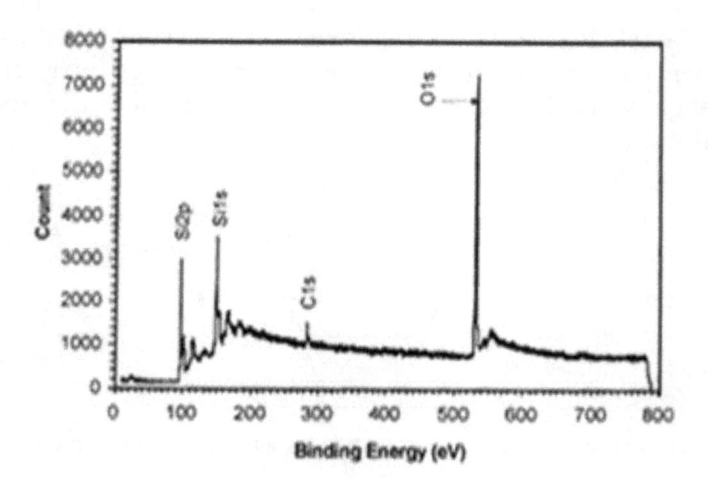


**Fig. 7.6.** XPS spectrum of the silicon surface exposed to ambient air.

The sensitivity of XPS can be increased by controlling the angle of the incident X-rays in the technique known as angle-resolved XPS (ARXPS). By varying the emission angle at which the electrons are collected, the detection of electrons from different depths is possible.

All in all, XPS in its basic configuration is a non-destructive surface characterization technique which is highly sensitive to changes in the chemical makeup of the surface. As such, it is routinely used in semiconductor device process development and diagnostics.

**Total reflection X-ray fluorescence (TXRF) spectroscopy.** An alternative to XPS, the method of X-ray fluorescence (XRF) spectroscopy used in material characterization is based on the emission of secondary (fluorescent) X-rays from the material that has been excited by exposure to high-energy X-rays. With incident X-rays' deep penetration and no escape depth limitation on secondary X-rays, information obtained in the case of XRF comes from the volume of the sample and thus does not provide information regarding the chemical composition of its surface. However, with the strong dependence of fluorescence signal on the angle at which primary X-rays irradiate the surface, the XRF spectroscopy can be made very sensitive to surface characteristics.

TXRF spectroscopy is a practical manifestation of this concept (Klokenkämper and Von Bohlen, 2015). In TXRF, the X-ray beam is directed at the surface at the glazing angle of less than 1°. As a result, only fluorescent X-rays emitted from the surface region reach the detector which identifies elements exposed to incident X-rays based on the energy of secondary X-rays (Fig. 7.7).

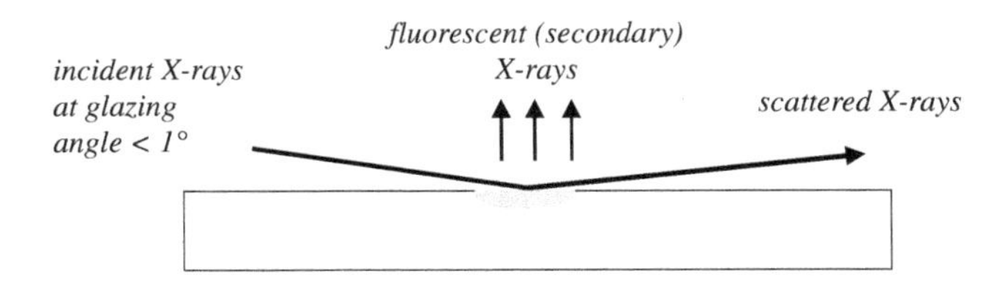


**Fig. 7.7.** Working of Total Reflection X-ray Fluorescence (TXRF) spectroscopy.

Limiting the response to the surface only is possible due to the negligible penetration of the near-surface region by the X-rays irradiating the surface at the glazing angle. As a result, the TXRF technique provides highly sensitive, non-destructive analysis of the chemical composition of the surface which in semiconductor practice is used primarily to detect metallic species acting as contaminants of the processed surfaces.

Together with XPS, the TXRF technique is a benchmark method in the characterization of semiconductor surface cleaning and conditioning applications in process development and diagnostics. In terms of applications in surface characterization, the difference between these two techniques is that by virtue of analyzing X-ray fluorescence rather than the energy of emitted electrons, the TXRF technique, unlike XPS, does not need to be carried out in a vacuum environment.

**Auger Electron Spectroscopy (AES).** The AES technique is similar to the XPS surface-specific technique of material characterization based on electron spectroscopy. The method exploits the Auger process (Pierre Auger, French physicist, 1899–1993), also referred to as the Auger effect, involving double ionization of multi-electron atoms. This means that, in contrast to XPS in which the emission of an electron carrying information about its origin involves one-step ionization, in the case of AES, ionization is a two-step process.

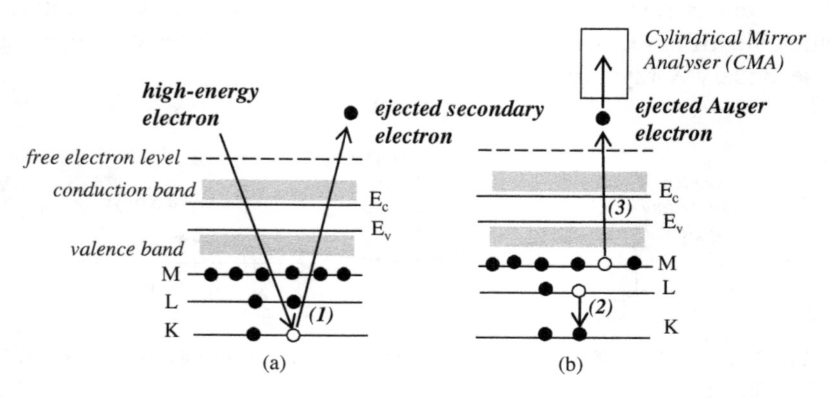


**Fig. 7.8.** Sequence in Auger process being a foundation of AES involves (a) impact ionization (1) and (b) ejection of Auger electron as a result of double-ionization process based on transitions (2) and (3).

The principle of the Auger process is schematically illustrated in Fig. 7.8. Depending on the way it is implemented and for what purpose, AES uses either beam of electrons or X-rays to stimulate desired intra-band interactions initiating the Auger process. In both cases, primary ionization causes the emission of secondary electrons from the deep level just like in XPS (transition (1) in Fig. 7.8(a)). In the Auger process, however, deep level vacated position is occupied by the electron from a higher level and transition (2) in Fig. 7.8(b) results in the release of energy. Released energy is used to stimulate the emission of the electron from the higher level to outside of the studied sample (transition (3) in Fig. 7.8(b)). The ejected electron is referred to as an Auger electron and the energy it carries, measured by a cylindrical mirror analyzer (CMA), determines its origin.

Due to the workings of the Auger process which involves three electrons in inter-orbital transitions, in the case of AES not only hydrogen featuring atomic number one, as in the case of XPS, but also helium with atomic number two cannot be detected. Other than that, AES provides accurate information regarding elemental composition in the top few nanometers of the sample (Ruzyllo *et al.*, 1976), while XPS is used to determine the surface chemical composition from somewhat larger analysis depths. When using a focused electron beam rather than X-ray excitation, AES allows better spatial resolution than XPS. On the other hand, the use of a focused electron beam may leave a local mark on the studied surface which is not the case when the surface of the sample is exposed to X-ray irradiation. To ensure undisturbed access of ejected electrons to the energy analyzer, Auger spectroscopy, similar to XPS, needs to be carried out in a high-vacuum environment.

As indicated in the above discussion, AES and XPS provide similar information regarding the chemical characteristics of the surface (Lindon *et al.*, 2017). From the perspective of the use for off-line process monitoring and diagnostics in semiconductor practice, the choice between AES and XPS depends on the needs regarding spatial and depth resolution of analysis as well as on equipment and expertise available.

**Electron microscopy.** Superior to optical microscopy in terms of resolution, electron microscopy is a well-established method, allowing the visualization of various features of the solid surfaces on micrometer and nanometer scales. From the perspective of discussion in this book, the limitation of scanning electron microscopy (SEM) is that for it to operate in a fully efficient fashion, scanned surfaces need to be electrically conductive. Otherwise, the emission of secondary electrons from the irradiated surface would not be sufficient to precisely determine the features of the surface. What it means is that surfaces investigated using electron microscopy need to be covered by a highly conductive material. In practice, most commonly, a thin layer of gold is deposited on the studied surface by, for instance, thermal evaporation. This requirement limits the usefulness of electron microscopy in the characterization of as-processed bare semiconductor surfaces which are of interest in the discussion in this book.

A powerful in terms of its capabilities variation of the electron microscopy is transmission electron microscopy (TEM) which in semiconductor engineering is most often used to study nanoscale multilayer material structures. As such, TEM is not a method that can provide information regarding stand-alone, as-processed semiconductor surfaces, and thus, it is not considered further in this book. On the other hand, however, the unique capability of TEM to display images of interfaces in multilayer material systems with atomic-scale resolution can be used to obtain information regarding the original condition of the surface upon which multi-layer structures are formed (see Fig. 1.7).

**Time-of-Flight Secondary Ion Mass Spectrometry (TOF SIMS or Static SIMS).** Proceeding with an overview of spectroscopic methods used in semiconductor surface characterization, we will now consider methods based on ion spectrometry. In contrast to electron spectroscopies in which the energy of electrons ejected from the solid is measured for the purpose of identification of elements present on the surface, in the case of ion spectroscopy, the mass of the ions sputtered off the surface and analyzed by mass spectrometers provides this information.

The most common semiconductor characterization method based on the determination of ion mass is secondary ion mass spectrometry (SIMS). Due to the high primary ion beam current, sputtering is involved in the process, and as a result, SIMS characterization features limited depth resolution and causes permanent damage to the analyzed material (Fig.7.9(a)). In addition, it annihilates some molecules, organic in particular, present on the exposed surface, thus preventing their detection. Consequently, SIMS, while being a technique that is broadly used in material characterization, in its original embodiment is not suitable for the characterization of semiconductor surfaces.

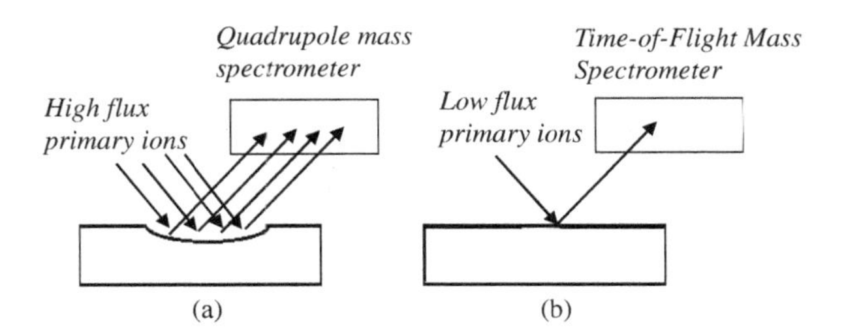


**Fig. 7.9.** Graphic representation of (a) dynamic SIMS and (b) static SIMS or TOFSIMS.

The damage-reducing, surface-sensitive version of the dynamic SIMS is a static SIMS also known as time-of-flight SIMS (TOFSIMS). As shown in Fig. 7.9(b), by using lower than conventional SIMS flux of primary ion beam, TOFSIMS reduces surface damage and improves the depth resolution of the SIMS method. In the case of conventional SIMS, also known as dynamic SIMS, the origin of high-energy ions ejected from the surface is determined based on mass spectroscopy.

In the case of time-of-flight SIMS (static SIMS), the mass of the secondary ions ejected, and thus their origin, is determined by measuring time-of-flight ions ejected from the surface need to reach mass spectrometer which determines their origin. This way TOFSIMS represents the variation of the SIMS method which allows the determination of some aspects of the chemical composition of the surface under investigation.

## 7.4 Ellipsometry

In agreement with conclusions drawn based on the review of optical phenomena occurring at the semiconductor surface (Fig. 7.3), the follow-up discussion of methods of optical characterization is focused on ellipsometry as a technique involving reflection of light which is compatible with monitoring of semiconductor surface condition.

The term "ellipsometry" refers to the technique which measures the thickness of transparent to light thin films and under proper conditions may identify selected features of the semiconductor surface underneath. Ellipsometric measurements provide information regarding the material system based on the change in polarization of an incident beam of light reflected off the sample surface (Fig. 7.10).

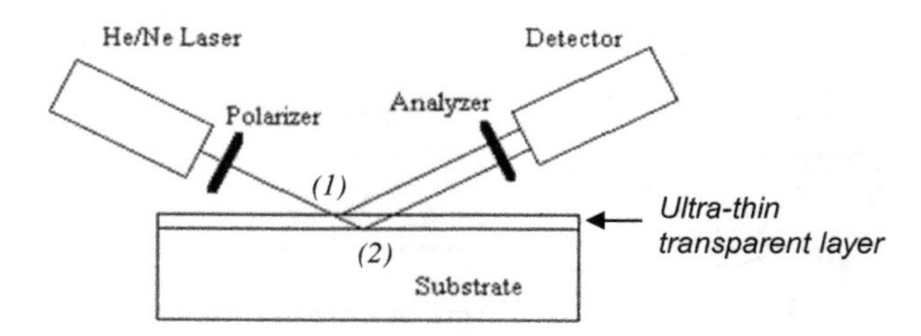


**Fig. 7.10.** Schematic illustration of the operation of a single wavelength ellipsometer.

Considering its versatility, ellipsometry is a technique broadly used in semiconductor research and engineering (Tompkins and Irene, 2005), including process monitoring applications. It represents indirect metrology where physical characteristics of the measured sample are obtained by fitting ellipsometric angles measured to the models established for a given material system including surfaces (Ohira and Itakura, 1979).

As indicated in Fig. 7.10, ellipsometry detects the change in polarization of the short-wavelength light upon its reflection from the air–material interface (1) and material–substrate interface (2). Single-wavelength ellipsometer shown in this figure uses a monochromatic light source, typically a laser in the visible spectral region, for instance, the HeNe laser featuring a wavelength of 632.8 nm.

While being used in semiconductor engineering primarily to determine the thickness of the film on the semiconductor surface of interest, ellipsometry also allows the characterization of composition, crystallinity, roughness, and other material properties that could affect the optical response obtained from the illuminated sample. In the case of ultra-thin (thickness in the range of 1 nm) native oxide covering the surface, for instance, processed results of ellipsometric measurements obtained may be an adequate representation of the condition of the surface. This feature makes ellipsometry useful in applications involving the characterization of semiconductor surfaces, keeping in mind, however, that the same does not apply to bare semiconductor surfaces.

Increased sensitivity of ellipsometry to the changes of surface features is accomplished by using spectroscopic ellipsometry (Fujiwara, 2007) which is a multi-wavelength version of ellipsometry and which is sensitive to the presence of surface layers on the order of a fraction of nanometer. While single-wavelength ellipsometry fulfills its role effectively when it comes to the measurement of the thickness of transparent dielectric films on isotropic surfaces, spectroscopic ellipsometry allows the characterization of the broader range of surfaces in terms of anisotropy and composition.

In summary, ellipsometry is a well-established, non-contact, non-destructive, precise analytical technique allowing determination at the nanoscale of physical properties of semiconductor materials and, under special conditions, properties of their surface. For these reasons, ellipsometers are often integral parts of the process tools where they are used for in-line monitoring of deposition, etching, and cleaning of the wafer surface.

## 7.5 Atomic Force Microscopy

As discussed in Section 4.2 of this *Guide*, roughness is a physical property of single-crystal semiconductor surfaces with an effect on the manufacturing yield and performance of devices formed on such surfaces. It is therefore essential that surface roughness is thoroughly controlled throughout the device manufacturing process. The method of atomic force microscopy (AFM) introduced earlier in this chapter is the method of choice when there is a need for the measurement and visualization of atomic-scale surface roughness. This section provides additional information on its operation and features.

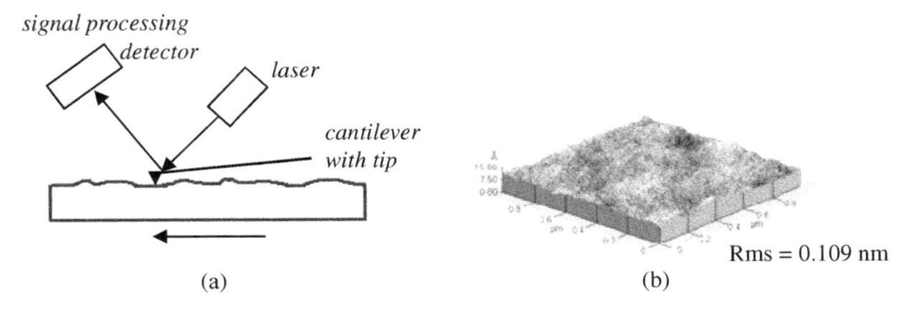


**Fig. 7.11.** (a) Schematic illustration of atomic force microscopy (AFM) setup and (b) AFM image of the surface of (111) silicon wafer used in the manufacture of devices.

Principles of AFM are schematically illustrated in Fig. 7.11(a) using as an example measurement setup operating in the most often used static mode, also known as the contact mode. The AFM uses ultra-sharp, atomic-scale processed probe which is attached to a small cantilever.

When the tip of the probe contacts the surface and is dragged across peaks and valleys reflecting surface roughness, the shape of the cantilever changes accordingly. This in turn alters the reflection of the laser beam off the cantilever which is subsequently detected by a photodetector. Based on the extent of bending of the cantilever scanning the surface, an image representing surface roughness is obtained (Fig. 7.11(b)) (Le Roux *et al.*, 2004).

As considered earlier in this chapter, the AFM method is based on attractive and repulsive atomic forces between two solids (probe attached to the cantilever and characterized material in Fig. 7.11) in close contact such as Van der Waals, electrostatic, and chemical forces (Haugstad, 2012). The change in the cantilever's bending reflects the tip–surface interaction force. Since the strength of the force decreases rapidly with distance, it is measurable only when the distance between the tip and the surface is within the 0.1–100 nm range. The nature of interactions between the probe and the surface is also dependent upon the chemical composition of the materials involved.

The AFM method is being implemented in various ways with the static mode shown above, or the contact mode, being the most common. Two other modes of AFM's operation which fall into the dynamic mode category are the non-contact and tapping modes. In the former case, the cantilever oscillates at its resonance frequency in the proximity of the surface of a sample. In the case of tapping mode, the probe comes into close contact with the surface intermittently.

The AFM characterization can be carried out in air, vacuum, and in some situations also in liquids, although, the highest resolution AFM images of atomic structures typically require a vacuum environment. No special preparation of the surface prior to AFM characterization is needed, which makes this technique compatible with the needs of process monitoring. The AFM is currently the most used method to precisely measure and quantify surface roughness and as such is commonly available in semiconductor research laboratories and manufacturing facilities.

Other than the AFM method used in advanced semiconductor research for surface imaging at the atomic level, which in contrast to AFM can only reproduce images of conducting or semiconducting surfaces, is scanning tunneling microscopy (STM). In STM, the tunneling current flows between the single-atom tip scanned across the sample at a distance of a few nanometers from the surface. Since the tunneling current measured is exponentially dependent on the distance between the tip and geometrical features of the sample, three-dimensional images of the surface representing its roughness can be obtained when the tip of the probe remains unchanged in the predetermined position and the surface of the probe is moved underneath.

To prevent disturbances of the tunneling current flowing between the surface and ultra-sharp tip of the probe, STM measurements need to be carried in the high vacuum using stands featuring superior vibration control. Furthermore, STM characterization uses are limited to highly conductive materials featuring clean surfaces. For all these reasons, while useful in surface research, STM is not used in the routine monitoring of semiconductor surface processes applied in device manufacturing.

## 7.6 Wetting (Contact) Angle Measurements

As indicated in the introductory comments earlier in this chapter, measurements of the wetting angle, also referred to as a contact angle, is a surface characterization technique which is easy to implement yet able to instantaneously provide information regarding changes in the condition of the surface under examination. Using simple instrumentation in the ambient air, measurements of the wetting angle readily distinguish between surfaces featuring different surface energies, including the difference between hydrophilic (wetting angle close to 0°) and hydrophobic (wetting angle approaching 90°) surfaces shown in Fig. 4.2. By measuring the values of contact angle, determination regarding changes in the condition of semiconductor surface as well as reproducibility of these changes can be readily established (Kwok and Neuman, 1999).

A schematic diagram showing the basic setup used to measure wetting (contact) angle $\alpha$ is shown in Fig. 7.12. The measurements performed using commercial tools or self-made experimental setups carried out in the laboratory environment, use pure water dispensed from the calibrated syringe as a surface-wetting liquid.

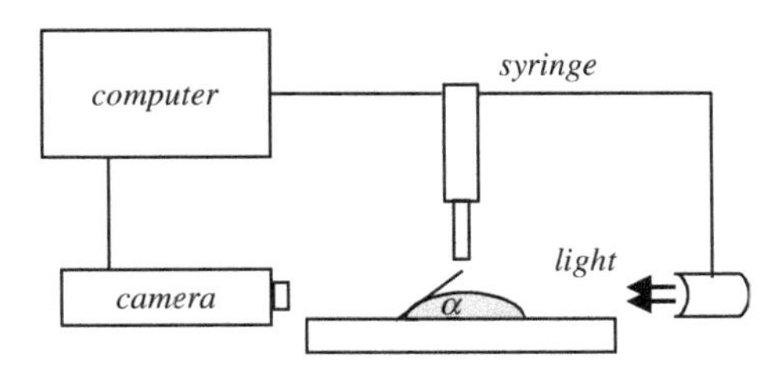


**Fig. 7.12.** Wetting (contact) angle $\alpha$ measurement setup.

An advantageous feature of the contact angle measurements is that they can be implemented on the as-processed surfaces without any prior preparatory treatments. This makes contact angle measurements suitable for inline process monitoring, although not on the product wafers (see the discussion in the following chapter).

## 7.7 Selected Methods of Electrical Characterization of the Semiconductor Surfaces

An introductory discussion concerned with electrical methods of semiconductor surface characterization in Section 7.1 considered techniques which provide information about the condition of semiconductor surfaces based on the interactions with electrically active centers on the surface and in the near-surface region. Information of this nature is particularly relevant as electrically active features of the surface have a direct impact on the operation of devices formed on such surfaces. Therefore, unlike techniques considered earlier in this chapter, electrical methods of surface characterization reveal properties of the material which directly impact the performance of semiconductor devices. For instance, while the causes of the reduced lifetime and mobility of charge carriers in the near-surface region of the processed semiconductor wafer may not be immediately known, and may require additional characterization using other methods, an adverse impact of this effect on the performance of the device such as MOSFET is well established. For that reason, electrical characterization methods providing direct quantitative information about the condition of the surface and near-surface region play an important role in semiconductor device engineering where the relatively minor uncontrolled changes in process parameters bring about readily measurable changes in the electrical properties of semiconductor surfaces.

The discussion in this section is limited to the review of selected methods of electrical characterization of semiconductor surfaces and near-surface regions which are compatible with the needs of measurements performed on the as-processed bare semiconductor surfaces. Furthermore, the focus of the follow-up discussion is on the in-line and on-line monitoring implemented in semiconductor device manufacturing. This consideration excludes from the discussion in this chapter electrical methods requiring the formation of permanent contact, such as metal–oxide–semiconductor (MOS) capacitor and Schottky diode-based characterization techniques as well as methods which require additional processing such as, for instance, annealing step aimed at the activation of structural defects or contaminants (Lagowski *et al.*, 1993). All these techniques play an important role in various applications in semiconductor device engineering but do not meet the requirements of process monitoring operations performed on the as-processed bare semiconductor surfaces on which the discussion in this book is focused.

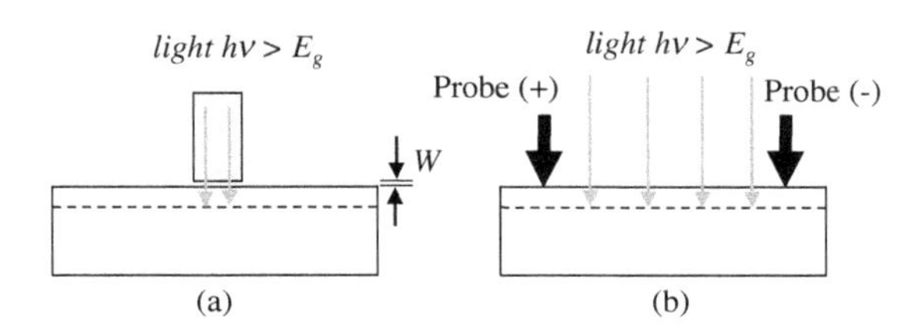


**Fig. 7.13.** Electrical methods of semiconductor surface characterization based on (a) non-contact and (b) lateral temporary contact measurements.

Recognizing the availability of a broad range of methods allowing the electrical characterization of semiconductor surfaces and interfaces, examples of the methods meeting the requirements listed above are further considered in the following. Specifically, the non-contact (Fig. 7.13(a)) and lateral temporary contact (Fig. 7.13(b)) methods of electrical characterization of semiconductor surfaces and near-surface region based on surface photovoltage (SPV) and photoconductance decay (PCD) effects, respectively, are considered.

**Non-contact method based on surface photovoltage (SPV).** As emphasized earlier, the discussion of non-contact semiconductor surface characterization methods, where the term "non-contact" means no physical contact of the electrically active parts to the measured surface, is limited to the methods which do not require any additional treatment such as annealing or oxidation prior to measurement. Furthermore, the measuring technique to be considered for process monitoring should not only be non-destructive but also non-invasive which means it should not alter the condition of the measured surface by mechanical contact, high energy irradiation, or high-density current flow across the surface. The same requirement also applies to otherwise non-contact methods such as those involving corona charging, for instance. Often employed in various surface characterization applications, corona charging by leaving the electric charge on the measured surface, which needs to be neutralized using additional treatment, is not compatible with the needs of process monitoring applications carried out on the as-processed surfaces. Also, scanning Kelvin probe microscopy, while being very effective in creating images representing selected features of surfaces under investigation (Marinskiy *et al.*, 2016), is of limited use in in-line semiconductor process monitoring on product wafers.

Among various methods of non-contact electrical characterization of semiconductor surfaces, a group of methods based on the effect of SPV meets the above-presented requirements (Roman, Brubaker *et al.*, 1998). Therefore, the follow-up simplified qualitative discussion of non-contact techniques of semiconductor surface characterization in this section exemplifies this class of methods.

In semiconductor engineering terms, *photovoltage* (PV) refers to the potential difference between contacts on the front and back surfaces of a semiconductor sample under illumination with light featuring energy higher than its energy gap and penetrating the illuminated semiconductor material. Using solar cells terminology, photovoltage is a voltage generated in the semiconductor material by the photovoltaic effect. The term "surface photovoltage (SPV)" refers specifically to the potential difference generated in the near-surface region of a semiconductor sample. Responsible for the initiation of SPV are generation–recombination processes stimulated by light absorbed in the near-surface depletion region ($W_d$) of the semiconductor sample. Due to the short wavelength used, for instance, 450 nm at which light is absorbed close to the surface (absorption depth in silicon is about 0.4 $\mu$m at this wavelength), the contribution of the bulk effects to SPV is minimized.

The absorption of light in the depletion region creates electron–hole pairs that separate due to the electric field in the near-surface space charge $Q_{sc}$ region (Fig. 7.14). Resulting minority carriers, electrons in this case, drift toward the surface where they recombine. To accomplish charge neutrality ($Q_s$=$Q_{sc}$), a measurable change in the voltage drop across the illuminated sample ($\Delta Vs$ in Fig. 7.14), corresponding to surface photovoltage is taking place.

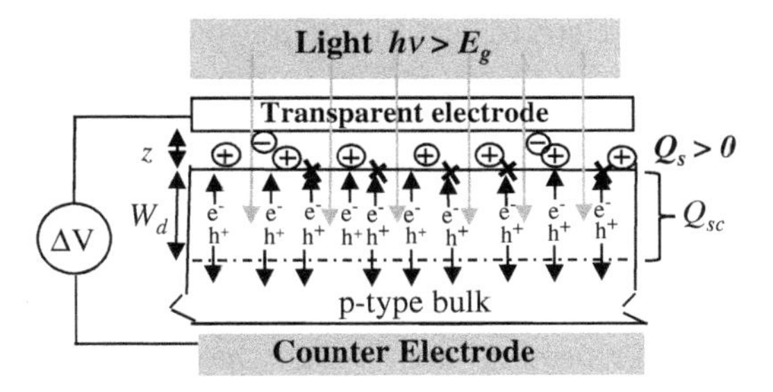


**Fig. 7.14.** Illustration of physical interactions during SPV-based non-contact surface charge measurement.

As mentioned earlier, the detection of the characteristics of the surface and near-surface region takes place in this case *via* capacitive coupling over very thin air gap (in the range of 20 $\mu$m) between the transparent electrode and the sample. In some cases, the electrically neutral and transparent-to-light spacer is used to ensure reproducible distance between the probe and the wafer.

Based on the measured value of $\Delta Vs$, the width of the space charge region $W_d$ and surface recombination lifetime can be determined using appropriate physical formulas (Kamieniecki *et al.*, 1994, 2001). Under depletion conditions, the total surface charge density can be calculated as $Q_s = N_{sc} W_{d,}$ provided that dopant concentration $N_{sc}$ in the space charge region is known, which is normally the case. Depending on the dopant concentration, the thickness of the space charge region $W_d$ varies from few micrometers in the case of lightly doped Si wafers to about 0.1 $\mu$m in the case of heavily doped wafers.

As it is apparent from this general description, the workings of the non-contact SPV methods in many aspects correspond to the procedures involving MOS capacitor test structures commonly employed in the characterization of semiconductor–oxide interfaces and semiconductor surfaces. The difference is in the air gap in the former case playing a role similar to the layer of oxide in MOS configuration.

From the point of view of process monitoring, the determination of $Q_s$ is the goal of the above-considered procedures. As discussed in Section 4.5, surface charge $Q_s$ represents the combined effect of charges related to electrically active surface states associated with disruptions of crystallographic structure at the surface of the measured samples (charge

$Q_1$ in Fig. 4.7(a)) and charges resulting from the interactions of the surfaces under evaluation with an ambient (charge $Q_2$ in Fig. 4.7(b)). In combination, measured surface charge $Q_s = Q_1 + Q_2$ is an adequate representation of the condition of the semiconductor surface subjected to various processes in the course of device manufacturing. It is also a good measure of stability and reproducibility of its characteristics from run to run.

The apparatus performing non-contact measurements discussed is schematically illustrated in Fig. 7.15. It consists of a motorized head probe which houses light source and electrode transparent to light and which can be moved up to load the wafer underneath and then down to perform the measurement. The wafer holder with counter electrode remains stationary while performing surface charge imaging (SCI) measurements or can be moved horizontally with respect to the upper part to implement the surface charge profiling (SCP) option. This last solution is a variation of SPV-based methods which by performing several measurements at various points on the surface of the same wafer allows the determination of the distribution of surface charge density over the area of the wafer.

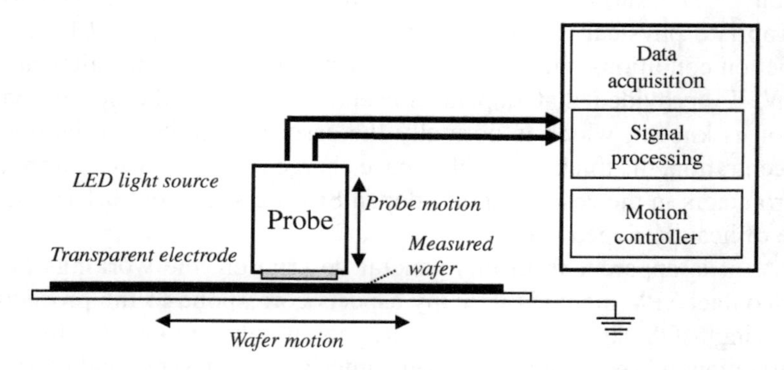


**Fig. 7.15.** Experimental setup designed to perform non-contact measurements of electric charge on a semiconductor surface.

The sensitivity of the SCP method to the varied conditions of the bare silicon surface is demonstrated in Fig. 7.16 which shows surface charge on the SC-1 treated silicon wafer, part of which was then immersed in $HF:H_2O$ solution. As demonstrated in Fig. 7.16, the SCP-measured surface charge responds very clearly to the varied conditions of the silicon surface. The low surface charge following SC-1 treatment results from metallic contaminants, mainly iron and aluminum, plated on the surface out of the SC-1 solution, while the higher charge on the HF:$H_2O$-treated part of the wafer results from surface protonization and reflects a predominantly hydrogen-terminated silicon surface.

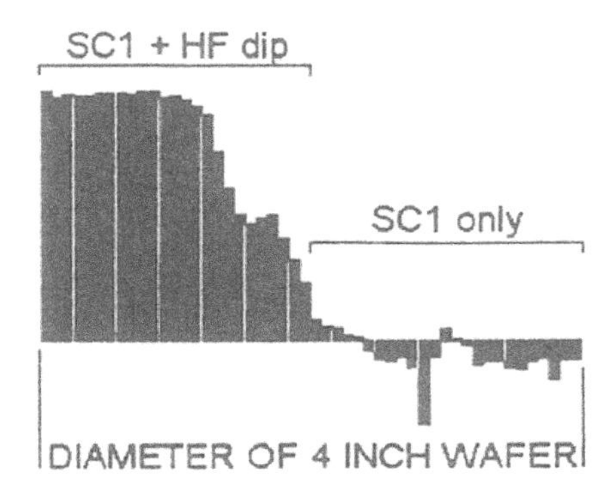


**Fig. 7.16.** Distribution of the surface charge on SC1+HF and SC-1-only treated surface of the same silicon wafer.

The effect of hydrogen termination on the value of surface charge is further demonstrated in Fig. 7.17, showing changes in the density of surface charge as a function of time of exposure to ambient air of the surfaces treated in 1% HF:$H_2O$ solution and anhydrous HF/methanol vapor. As seen in this figure, the former, featuring superior hydrogen termination of the silicon surface, ensures the stability of the surface condition for a significantly longer time as compared with the Si surface treated in AHF/ $CH_3OH$ vapor (Roman *et al.*, 1995).

From the perspective of practical applications of the measurements implemented using the setup shown in Fig. 7.15, of importance is the distance separating the probe from the measured surface. With separation, possibly controlled by the spacer, in the range of the tens of micrometers and assuming possible variations of wafer surface texture, unacceptable in process monitoring on product wafers, physical contact of the parts of the probe with the analyzed surface cannot be entirely prevented. Consequently, while useful in on-line process monitoring and

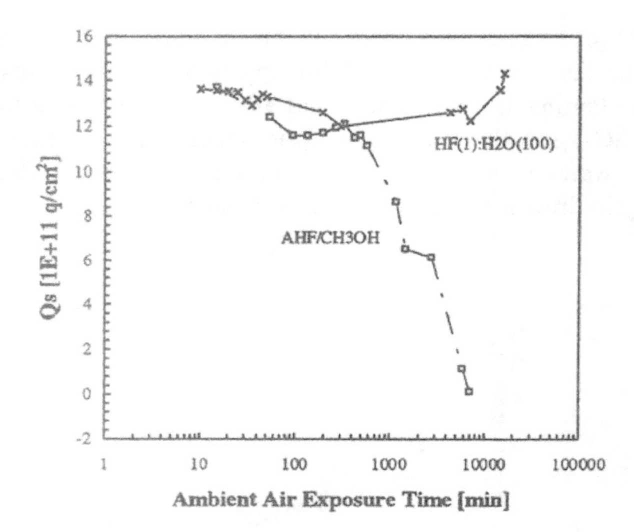


**Fig. 7.17.** Changes in the density of surface charge as a function of the time of exposure to dilute HF:water solution (HF(1):$H_2O$(100)) and anhydrous HF:methanol (AHF/$CH_3OH$) vapor.

diagnostics involving test wafers, the SPV-based method in the version considered in Fig. 7.15 may not be suitable for some applications involving in-line process monitoring on product wafers (see the discussion in Section 8.1).

A solution to this challenge is offered by a patented methodology involving SPV surface characterization in which coupling between the probe and the surface is modified by radio frequency (RF) signals (Kamieniecki, 2011). With this modification, measurements of surface charge, and then determination of the minority carrier lifetime and mobility in the near-surface region, can be carried out at the distance between the probe and the wafer surface (Fig. 7.15) in the range of a millimeters rather than micrometers. In this variety, a non-contact method based on SPV discussed here could be compatible with in-line process monitoring on product wafers.

As it can be concluded from this discussion, the non-contact method considered allows a reliable determination of the density of electric charge on the semiconductor surface from which information regarding the characteristics of the charge carriers in the near-surface region such as their lifetime and mobility can be derived. In terms of in-line process monitoring, compatible with process monitoring on product wafers, the RF version of the method can measure the density of surface charge adequately representing electronic properties of the surface and as such can be used in-line as a measure of process performance and reproducibility. Examples of experiments involving the measurement of surface charge using the SPV methodology are included in the discussion in Chapter 8.

**Temporary Contact Photoconductance Decay (TC-PCD) method.** In terms of surface characterization using temporary contacts, of interest here is a setup featuring lateral configuration with both ohmic contacts on the measured surface (Fig. 7.13(b)). Lateral configuration is chosen because in the case of vertical setup (ohmic contacts on the front and back surfaces of the sample, Fig. 7.4(b)), current flows across the bulk of the sample making the separation of surface-related effects from those controlling current flow in the bulk of the sample not a straightforward matter.

In the laterally configured test structures (Fig. 7.13(b)), current flows parallel to the surface in its immediate vicinity (Drummond *et al.*, 2009). As a result, it is strongly affected by its characteristics including structural defects acting as recombination–generation centers which affect the lifetime of charge carriers. Also, atoms displaced from their original lattice sides at the surface are interfering with the flow of the current by altering the mobility of charge carriers. In short, the lateral configuration of the test structure, combined with excitation using light featuring shallow absorption depth in the semiconductor sample, allows effective separation of the surface-related effects from those controlling bulk characteristics of the illuminated material. For these reasons, the discussion of methods based on temporary contact in this section is focused on the lateral configuration shown in Fig. 7.13(b).

The temporary contact photoconductance decay (TC-PCD) method in lateral configuration allows direct identification of surface structure defects acting as minority carrier recombination centers and contributing to the surface charge $Q_1$ shown in Fig. 4.7. In addition, based on the PCD results, minority carrier mobility in the near-surface region can be calculated to provide information regarding structural defects caused by the abrupt discontinuity of the lattice at the surface and acting as scattering centers for the moving carriers.

Figure 7.18 illustrates physical interactions in the near-surface region of the semiconductor sample initiated by illumination with light featuring energy $h\nu$ higher than the energy gap $E_g$ of the illuminated semiconductor.

With probes forming temporary ohmic contacts to the surface as shown in the figure, exposure of the surface to the pulsed light, rather than continuous illumination, allows measurements of parameters reflecting the condition of the near-surface region of the semiconductor sample (Drummond, Kshirsagar *et al.*, 2011). Considering pulsed light-induced near-surface effects, it is assumed that surface charge $Q_s > 0$ (Fig. 7.18) originates solely from the positively charged centers associated with the disrupted crystallographic structure of the semiconductor sample at the surface and is independent of the ambient in which measurements are carried out.

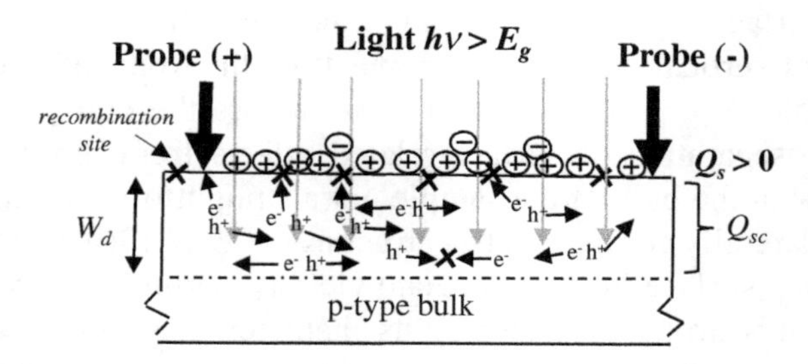


**Fig. 7.18.** Illustration of physical interactions during PCD-based lateral temporary contact measurement performed on a semiconductor wafer.

The way conductance, *G*, defined as *1/R* where *R* represents electrical resistance of the portion of material under illumination between two contacts, responds to the pulsed illumination, is a foundation upon which quantitative information regarding physical characteristics of the near-surface region can be obtained using this method. Responsible for the increase of conductance in the near-surface region are electron–hole pairs generated within the optical absorption depth underneath the surface. With voltage applied to the probes, electrons and holes are moving in the opposite direction and eventually recombine contributing to the decay of photoconductance. Electrons photo-generated in the immediate vicinity of the surface are swept by the electric field associated with surface charge $Q_s$ toward the surface where they recombine coming across the structural defects acting as recombination sites also contributing to the photoconductance decay. At the same time, photo-generated holes are driven by the electric field toward the bulk of the substrate away from the surface (Fig. 7.18). Surface recombination dominates the carrier lifetime when the surface defect density is very high, for instance, due to surface micro-roughness resulting from invasive interactions with chemically and physically active species during surface cleaning or etching operations.

Interpretation of the origin and location of recombination centers within semiconductor surface and near-surface region based on photo-conductance decay is a function of many variables and requires careful analysis details of which are escaping simplified interpretation. What is important in this discussion, however, is that analysis of the time-resolved photoconductance decay following a short pulse illumination with energy $E_{hv}>E_g$ is a useful method for determining carrier recombination lifetime in the illuminated near-surface part of the semiconductor. This in turn allows detection of recombination centers associated with structural and compositional imperfections of the lattice, specifically in this region.

As the temporary contacts to the surface, sharp point contacts made out of mechanically and chemically stable tungsten are used. Metal point contacts are preferred in this application over soft metal probes, mercury for instance, as the latter are getting readily contaminated by the residues collected from the measured surfaces and cannot assure long-term reproducibility of the measurements performed. Furthermore, sharp point contacts penetrate through the layer of residues possibly present on the measured surface allowing the formation of the contacts to the surface which without any additional surface treatments display characteristics of the ohmic contacts.

The PCD measurements discussed are carried out using the measurement setup shown in Fig. 7.19(a), in which an important role is played by the pulsed signal powering the laser diode which in turn brings about pulsed illumination of the wafer under investigation (Drummond *et al.*, 2009). In the case of the 658 nm laser diode often used in this application, a photon energy $hv$ of 1.88 eV exceeds the 1.1 eV energy gap of silicon ($E_{hv} > Eg$) at the light penetration depth in silicon of approximately 4 $\mu$m (Fig. 3.11).

Changes in the slope of the PCD trace during the photoconductance decay cycle are being used as an indication of the changes in the density of charge carriers in the near-surface region resulting from interactions with surface recombination centers. Figure 7.19(b) shows a short pulse illumination with energy $E_{hv} > E_g$ and corresponding photoconductance output traces reflecting changes in the defects acting as recombination centers in terms of their density and location.

Three photoconductance decay traces in Fig. 7.19(b) represent three different situations in this regard with the slope of each curve corresponding to the rate at which charge carriers are recombining. The curve (1)

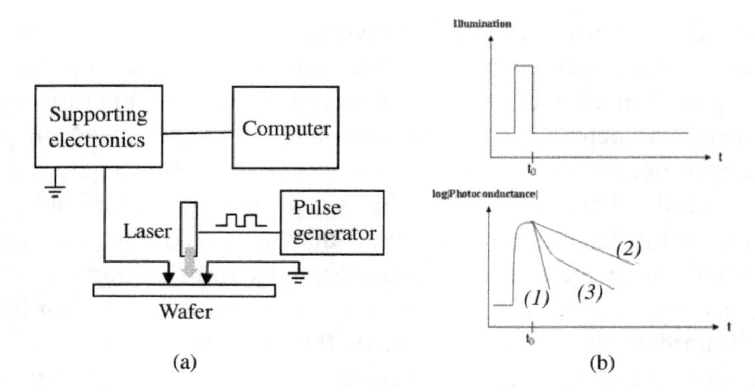


**Fig. 7.19.** (a) Experimental setup determining near-surface characteristics of semiconductor wafer based on PCD measurements and (b) dependence of the PCD signal on the condition of near-surface region.

featuring rapid decay of photoconductance indicates a severely damaged surface which promotes increased recombination resulting in a short minority carrier lifetime in this region. On the other hand, the shape of curve (2) in Fig. 7.19(b) indicates significantly slower photoconduction decay, minimal surface damage, and thus, longer minority carrier lifetime controlled by defects predominantly beyond the surface region. The shape of the PCD curve observed in many practical situations is represented by the trace (3) in Fig. 7.19(b). Here, the initial period of photoconductance decay indicates the impact of the recombination centers located near the surface. Initially faster decay of photoconductance occurs before the carriers photo-generated near the surface diffuse toward the bulk of the sample where a longer carrier lifetime is responsible for reduced photo-conductance decay.

A calculation of recombination lifetime from only the initial slope of the log of the photoconductance decay curve yields a minority carrier lifetime $\tau$ dominated by the properties of the near-surface region. As explained in the literature (Drummond, Bhatia *et al.*, 2011), it is based on the extraction of the value of $\tau$ from the time and carrier lifetime-dependent changes of the photo-generated excess carrier concentration $\Delta n(t) = \Delta n(0)\exp(-t/\tau)$. From the obtained PCD trace, one can determine the time required for the photoconductance to decay from the peak value to the time at which photoconductance has dropped to the level of *1/e* peak value. The measured time interval represents the value of $\tau$.

An example of experimental results obtained using temporary contact PCD characterization method on the wafers featuring different surface roughness is shown in Fig. 7.20 (Drumond, Bhatia *et al.*, 2011). Traces in this figure clearly distinguish between the wafers featuring surface roughness RMS of 1.04 nm and 0.32 nm featuring correspondingly longer minority carrier lifetime in the latter case. The shape of traces and the rate of photoconductance decay in Fig. 7.20 reflects the shape and rate of decay in the case of theoretical traces in Fig. 7.19.

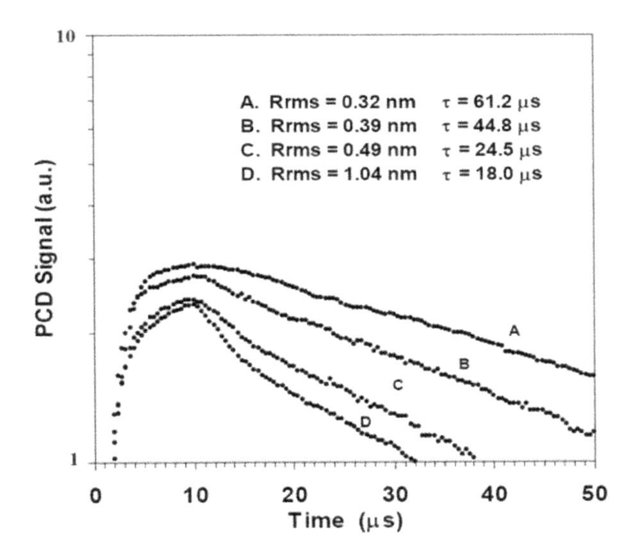


**Fig. 7.20.** PCD signal (arbitrary units) measured using temporary contact as a function of surface roughness.

The correlation between the shape of photoconductance decay traces and the condition of the near-surface region of the silicon wafer is further demonstrated in Fig. 7.21. The PCD traces in this figure were measured using temporary contacts on the front polished (trace *A*) and back unpolished surface of the same wafer (trace *B*). The former, due to the smoother front surface of the wafer, allows distinction between the initial shorter minority carrier in the near-surface region and the significantly longer one in the bulk of the sample.

In the latter case, trace B shows rapid decay of photoconductance indicating a higher density of defects acting as recombination centers in the near-surface region associated with the unpolished back surface.

As indicated in the above discussion, minority carrier lifetime can be determined directly from the PCD response. In contrast, the determination

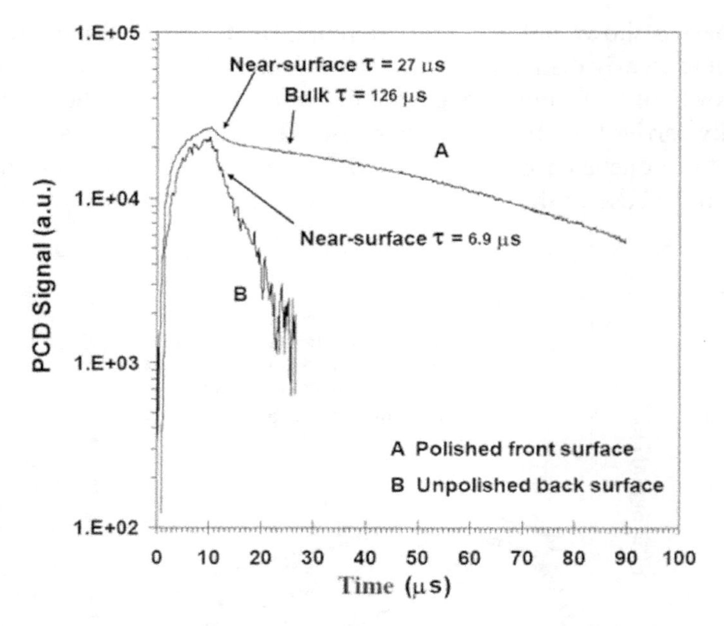


**Fig. 7.21.** PCD signal as a function of time measured using lateral temporary contact on the front (polished) and back (unpolished) surfaces of the same silicon wafer.

of the near-surface mobility of charge carriers in addition to the measured minority carrier lifetime requires knowledge of other quantities such as carrier generation rate and absorbed optical power. A possible solution, valid from the point of view of process monitoring, is to assume that those other inputs would remain constant from measurement to measurement so that the obtained mobility results would indicate relative changes in carrier mobility in response to changing surface conditions only. The validity of this assumption is demonstrated by the results shown earlier in Fig. 2.13, illustrating the changes in charge carrier mobility measured using the PCD technique discussed in this section as a function of surface roughness.

In agreement with the scope of this book, the review of semiconductor surface characterization methods in this chapter was focused on the methods which specifically determine the properties of the as-processed, bare semiconductor surfaces. In other words, methods of interest in this discussion were those which do not require any treatments of the wafer prior to measurement and therefore can be performed on the wafers immediately following any given processing step performed on the wafer surface.

# Chapter 8

# Characterization of Semiconductor Surfaces in Process Monitoring

## 8.1 Introduction

Following on the review of surface characterization methods used in semiconductor device technology in the previous chapter, the goal of this chapter is to consider the practical aspects of the application of surface characterization methods in the monitoring of processes used in the manufacture of semiconductor devices. In agreement with the scope of this book, the follow-up discussion is focused on the monitoring of bare silicon surfaces processed during front-end-of-line (FEOL) operations such as cleaning and etching applied in the device fabrication sequence.

This chapter considers the reasons driving the applications of semiconductor surface characterization methods in broadly understood semiconductor device technology. The observations herein are based on the author's experiences with semiconductor surface-related research and are formulated with a recognition of the fact that the challenges considered may be seen differently depending on the nature and purpose of the procedures considered.

To establish an understanding of the concepts underlying the considerations in this chapter, the goals of surface characterization in semiconductor device technology are examined first. Then, process monitoring modes and principles upon which semiconductor surface characterization can be used to assess the outcome of operations performed on the wafer during device manufacturing are reviewed. In the remaining sections of this chapter, the examples of specific surface characterization procedures compatible with the needs of process monitoring as understood in this discussion are considered.

## 8.2 Semiconductor Surface Characterization in Device Technology

The working of a semiconductor material system processed into a functional device depends on its interactions with stimuli from outside sources which can include electrical and magnetic fields, current, voltage, light, temperature, and mechanical stress. In many cases, the surfaces of semiconductor materials involved play a role in controlling these interactions. Therefore, in order to ensure predictable and reproducible characteristics of the device, the properties of the surfaces of semiconductor materials used to fabricate any given element must be known and controlled not only in terms of their inherent characteristics but also in terms of the changes inflicted by the processes involved in the manufacture of these devices. As a result, surface characterization is an integral part of the procedures involved in the manufacturing of semiconductor devices at the stages of the fabrication sequence when the surface remains exposed to outside interactions.

An issue of interest in this discussion is identification of applications in which surface characterization is being used at various stages of the development of any given semiconductor device technology. Figure 8.1 schematically illustrates areas in which surface characterization is a necessary component.

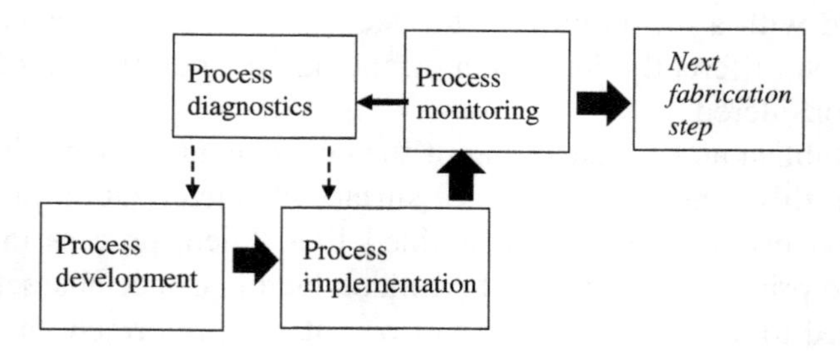


**Fig. 8.1.** Schematic representation of the areas in which surface characterization procedures are applied in semiconductor device technology.

**Process development.** Based on the established scientific foundation regarding surface characteristics of semiconductor material of interest, process development identified in Fig. 8.1 is concerned with surface processing and characterization methodology needed to convert a semiconductor material into a functional device by applying procedures and tools used in semiconductor manufacturing. Seen this way, process development is an extension of the basic surface research into the applied research carried out with a specific device and a specific set of tools in mind. At this stage, any surface characterization method from among those discussed in Chapter 7 is being applied as needed.

**Process implementation.** Once all the fabrication parameters are determined and the appropriate tools selected based on the outcome of the process development stage, the process is implemented in the device manufacturing environment. Depending on the nature of the process and the way it is carried out, surface characterization procedures can be either part of it or can be carried out by employing dedicated process monitoring procedures.

**Process monitoring.** It is essential to control the condition of the semiconductor surface under processing at every stage of the fabrication sequence during which the surface of the semiconductor wafer is exposed to external interactions. Therefore, the effect of semiconductor manufacturing processes on the condition of the substrate's surface must be monitored and any departure from the predetermined surface characteristics detected in real time using appropriate process monitoring techniques. Needs in this regard make surface characterization carried out for the purpose of process monitoring an integral part of the semiconductor device manufacturing sequence.

As shown in Fig. 8.1, the next step in the device fabrication sequence depends on the result of the process monitoring procedure. In the case where it indicates the correct performance of the monitored process, the wafer is transferred directly to the tool carrying out the next operation in the manufacturing sequence. In the case of a result indicating process malfunction of any nature, the wafer is subjected to process diagnostics aimed at the identification of its causes.

**Process diagnostics.** Any deviation from the predetermined surface characteristics unraveled during process monitoring is an indication of the process malfunction which needs to be diagnosed and corrected. Procedures used to identify the nature of uncontrolled changes in the surface condition detected during process monitoring, based on which the cause of the process failure can be identified, are referred to in this discussion as process diagnostics (Fig. 8.1).

Process diagnostics is not a part of the routine device manufacturing procedures unless monitoring of any given surface processing step indicates its malfunction. In such cases, the evaluation of the surface condition using methods appropriate to the needs is carried out. Then, as shown in Fig. 8.1, the results of the failure analysis are fed back to the process implementation phases which can be modified as needed. In the case, process diagnostics would unravel the reasons for the process malfunction warranting major modifications in the device fabrication sequence the results of process diagnostics need to be fed back to the process development stage.

Keeping in mind considerations related to surface engineering procedures applied in device manufacturing and presented in Fig. 8.1, the discussion in the remainder of this chapter is concerned with the process monitoring operations involving specifically surface characterization procedures.

## 8.3 Process Monitoring in Semiconductor Device Technology

Control of the outcome of individual manufacturing steps, or quality control, is an integral part of any industrial manufacturing endeavor including the fabrication of semiconductor devices. This, however, distinguishes itself among the others by involving processes manipulating matter at the atomic and molecular level, and thus, imposing particularly stringent requirements regarding techniques applied in the control of semiconductor device fabrication processes. Furthermore, the environment involving cleanrooms in which semiconductor device manufacturing takes place imposes additional requirements regarding the way quality control in semiconductor device manufacturing is carried out.

Depending on the features of the processes subjected to monitoring, and the purpose for which monitoring of surface condition is applied, the surface characterization processes implemented in-line, on-line, and off-line are distinguished in the follow-up discussion.

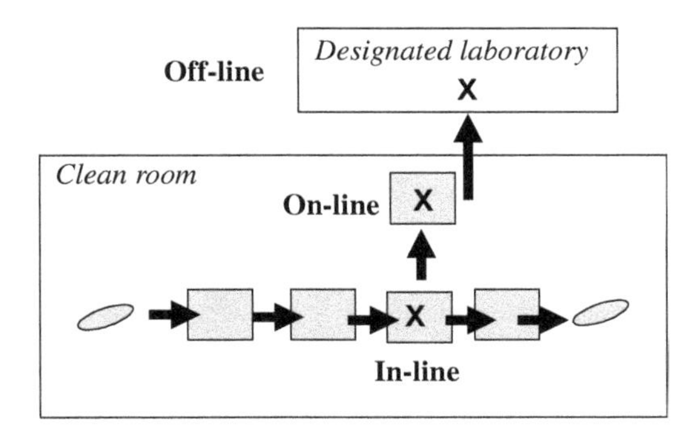


**Fig. 8.2.** Implementation of process monitoring procedures in semiconductor device manufacturing facilities.

Three different process characterization modes in semiconductor device manufacturing are illustrated in Fig. 8.2, showing simplified schematics of the clean room with the part of the manufacturing line identified. Symbol "*X*" in this figure denotes points at which operations involving surface characterization are carried out for the purposes of in-line and on-line process monitoring as well as off-line process diagnostics.

Among them, of interest in this discussion are in-line and on-line surface characterization procedures which can be used to monitor the results of the treatments to which the semiconductor surface of interest is subjected including cleaning and etching. Not covered in this discussion are measurements carried out off-line which, while indispensable in process development and diagnostics, are not involved in real-time process monitoring.

**In-line process monitoring.** This monitoring mode is of special interest in this discussion which is focused on the characterization of bare semiconductor surfaces carried out to detect possible process malfunctions. The way in-line process monitoring is understood is explained using schematics in Fig. 8.3 for a case of single wafer processing. The goal is to determine the performance of Process A, representing surface treatment such as cleaning and conditioning prior to the wafer transfer to chamber B carrying out, for instance, thin-film deposition.

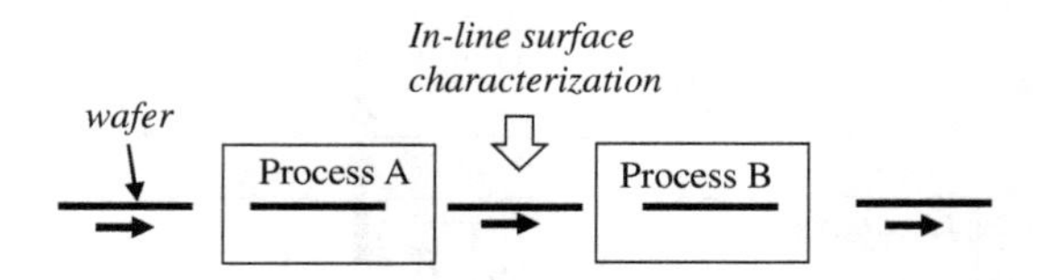


**Fig. 8.3.** In-line process monitoring on product wafers.

To make the in-line process monitoring scenario represented in Fig. 8.3 feasible, a surface characterization procedure compatible with in-line process monitoring should involve a method which does not require mechanical contact between the surface and measurement apparatus and, thus, is contactless and non-invasive. This last requirement means that the measurement should not affect in any way, through chemical or physical interactions, either the wafer surface or the ambient in which measurement is performed. Furthermore, the in-line process monitoring method should provide quantitative information related to surface conditions, such as, for instance, the density of surface charge, in real time and based on the measurements of product wafers rather than on designated test wafers.

Only some methods of semiconductor surface characterization can be adapted to the specific needs of in-line process monitoring as defined above. Later in this chapter, surface characterization methods potentially compatible with the needs of this demanding application are considered.

**On-line process monitoring.** In general, on-line process monitoring involves procedures which, unlike in-line monitoring, are not integrated in the manufacturing line, yet include instrumentation installed and used in the clean room manufacturing facility in the vicinity of such line (Fig. 8.2). This approach, schematically illustrated in Fig. 8.4, is mostly associated with batch processes and represents the traditional way various process monitoring tasks are implemented in semiconductor research laboratories and production facilities.

As shown in Fig. 8.4, on-line process monitoring is performed on the designated test wafers rather than on product wafers subjected to in-line process monitoring (Figs. 8.2 and 8.3). In contrast, methods used in

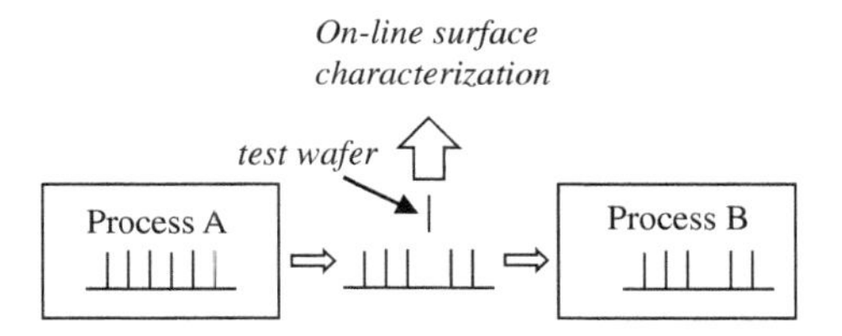


**Fig. 8.4.** On-line process monitoring in the case of batch processes is performed on the designated test wafers.

on-line process monitoring typically affect the condition of the monitored surface, and thus, test wafers subjected to on-line testing are not used in the fabrication of final devices. In addition, on-line characterization often requires time-consuming operations and complex instrumentation which makes it incompatible with real-time process monitoring.

Following the above comments on in-line and on-line process monitoring, later in this chapter, examples of the methods compatible with both these applications are considered. The general rules underlying the concept of process monitoring based on the characterization of semiconductor surfaces are also considered.

As indicated in the discussion in this book, the control of the outcome of various operations performed on semiconductor wafers is an integral part of semiconductor device manufacturing technology. The process monitoring objectives, methods used, monitored materials, as well as procedures and instrumentation employed vary depending on the nature of operations to which semiconductor wafers are subjected and at what stage of device fabrication sequence monitoring is carried out. Consequently, different process monitoring methodologies are employed early in the device fabrication sequence (front-end-of-line, FEOL, process) and late in the device fabrication process (back-end-of-line, BEOL, process).

Surface cleaning and conditioning operations considered here, as well as etching processes exposing the surface carried out during the FEOL stage of the manufacturing procedures are routinely subjected to process monitoring. Whether it is a surface of the wafer without any pattern (Fig. 8.5(a)) or a bare silicon surface on the partially patterned wafer (Fig. 8.5(b)), direct access to the monitored surface is a pre-condition for its successful characterization. During the later stages of the fabrication sequence involving back-end-of-line (BEOL) processes, the surface of the semiconductor substrate wafer is covered with layers of various mostly non-semiconductor materials (Fig. 8.5(c)) and monitoring of its condition is not possible and not needed. Therefore, surface monitoring methods and methodologies discussed here are concerned solely with FEOL processes.

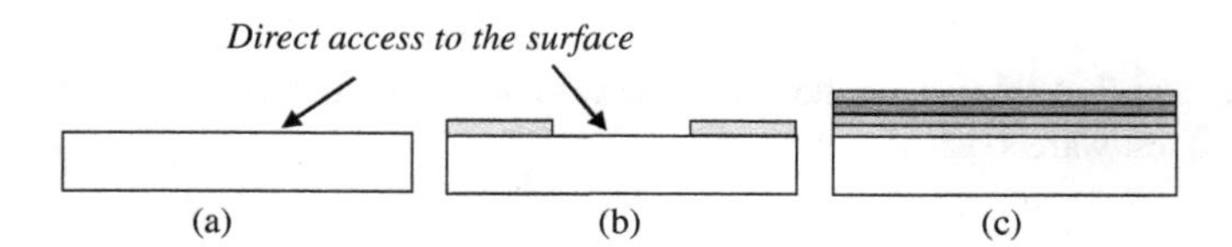


**Fig. 8.5.** (a) In the case of FEOL processes, the bare semiconductor surface is exposed and available to monitoring, (b) partially exposed allowing monitoring of the surface condition following etching, and (c) at the BEOL stage of the fabrication sequence, the semiconductor surface is covered with other materials and not accessible to monitoring its condition.

In the remaining discussion in this chapter, the selection of surface characterization methods which are compatible with process monitoring of as-processed bare semiconductor surfaces is considered.

## 8.4 In-line Process Monitoring Based on Characterization of Semiconductor Surfaces

The goal of the discussion in this section is to consider examples of the ways in-line process monitoring based on the determination of the condition of semiconductor surface is being implemented in the manufacture of semiconductor devices. Selection of the methods to be used for the specific process monitoring application depends on the type of process and conditions under which it is carried out. As indicated, of interest are methods that provide information regarding the properties of the as-processed bare semiconductor surface and near-surface region and are suitable for real-time process monitoring applications.

There are various ways process monitoring procedures can be applied in semiconductor device manufacturing. The viable options in this regard are offered by the process integration using cluster tools which allow the implementation of the multistep fabrication sequence without exposing the wafer to the ambient air. As an example, considered here is a solution involving a four-sided cluster tool schematically illustrated in Fig. 8.6

designed to implement thin-film dielectric or metallic compound deposition sequence (Lee *et al.*, 2002). It is assumed that the condition of the surface of the silicon wafer acting as a substrate influences the characteristics of the resulting structure and, thus, must be treated prior to thin-film deposition. The outcome of the surface treatment step needs to meet predetermined specifications which should be verifiable by the designated monitoring step prior to the deposition process.

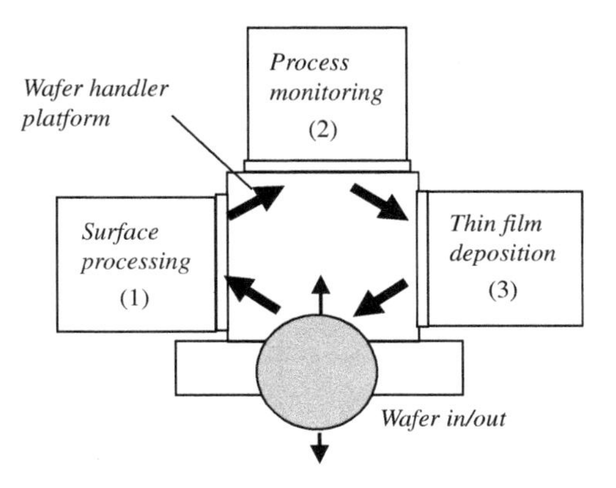


**Fig. 8.6.** Four-sided cluster tool allowing in-line process monitoring on product wafers.

As shown in Fig. 8.6, the cluster is constructed around a wafer handler platform working at the reduced pressure or in the atmosphere of an inert gas. The handler is equipped with a robotic arm loading and unloading wafers into reactors designed to perform specific functions. The pressure at which each reactor is operating, the composition of gaseous ambient used, as well as the wafer temperature are independently controlled in the reactors installed on the cluster.

Loaded into the cluster, the wafer is first transferred to the reactor (1) where it is subjected to surface treatment carried out in the gas phase as needed. In order to ensure that the surface of the wafer is properly prepared for the thin-film deposition step in the reactor (3), the wafer is transferred by the robotic handler to chamber (2) to perform the evaluation of the surface condition, applying procedures considered in this discussion as an in-line process monitoring step (Roman, 2001).

At the end of the sequence carried out in the cluster shown in Fig. 8.6, the wafer is transferred to the reactor (3) where the thin-film deposition process is carried out. What is important from the surface interactions point of view is that all the operations listed are performed without exposing the surface of the wafer to ambient air.

An alternative approach to surface characterization for the purpose of process monitoring applies to the processes in which process monitoring function is carried out *in situ*. A process which allows the integration of process monitoring and film deposition functions into a single reactor is the process of molecular beam epitaxy (MBE) which on one hand is among the most demanding in terms of pre-deposition characteristics of the substrate's surface (Ishizaka and Sharaki, 1986), while on the other provides a unique environment for the implementation of the in-line process monitoring procedures. Figure 8.7 shows schematics of the MBE reactor used for epitaxial deposition of semiconductor elemental and compound materials with atomic-scale precision.

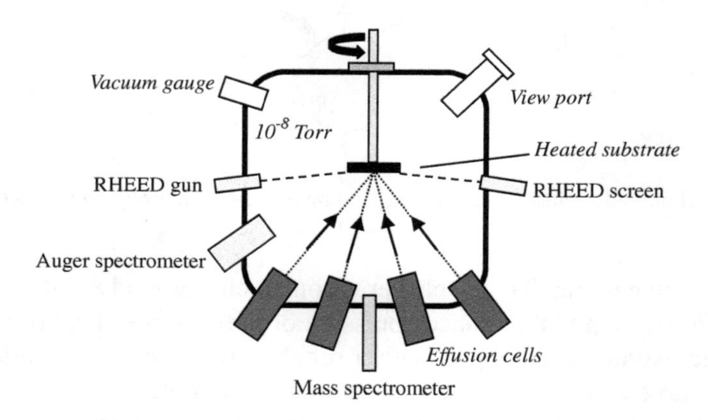


**Fig. 8.7.** Schematic diagram of the reactor performing MBE illustrating in-line process monitoring options.

The MBE processes are carried out under the ultra-high vacuum ($10^{-8}$ Torr and lower) and at the substrate temperature typically in the range of 600–800°C. The effusion cells (Fig. 8.7) generate components of the deposited films which are then delivered to the heated substrate. In all this, the advantage is being taken of ultra-high vacuum, under which deposition is carried out. This feature of the environment in which MBE processes are implemented determines the selection of methods used to monitor MBE deposition.

As the name indicates, the MBE process is concerned with epitaxial deposition and calls for process control solutions able to follow the growth of the film requiring precise alignment of atomic layers. In the case of MBE, the reflection high-energy electron diffraction (RHEED) method considered in Section 7.2 is commonly used for this purpose (Fig. 8.7). In addition, as indicated in the figure, to monitor chemical composition of the epitaxially grown films, the MBE reactors can be equipped with either quadrupole mass spectrometer or Auger spectrometer also considered in the previous chapter. On the other hand, however, other than ultra-high vacuum annealing causing desorption of surface-contaminating molecules, MBE technology does not provide diverse surface processing options.

Similar to MBE, the methods of atomic layer deposition (ALD) are compatible with *in situ* monitoring of the surface condition prior to the deposition of non-crystalline films. Often used for this purpose is spectroscopic ellipsometry considered briefly in Section 7.3. Figure 8.8 illustrates the integration of a spectroscopic ellipsometer with a plasma-enhanced ALD tool used to deposit thin-film dielectrics.

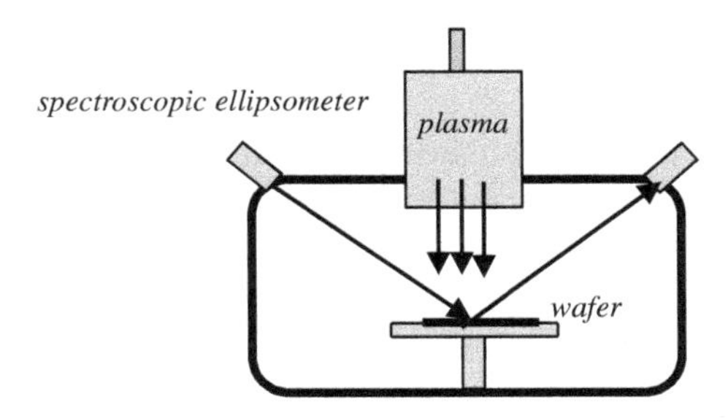


**Fig. 8.8.** Spectroscopic ellipsometer integrated with plasma-enhanced ALD tool for the purpose of monitoring the condition of the wafer prior to thin-film deposition.

The approaches to in-line process monitoring applied in semiconductor device manufacturing considered in Figs. 8.6–8.8 involve surface characterization methods, accommodating specific needs of the process and design of the tools used. Other than that, there is a range of surface characterization methods potentially of use in in-line and on-line monitoring of manufacturing procedures not subjected to the constraints imposed by the design of the reactors involved.

## 8.5 Methods Compatible with In-line and On-line Process Monitoring of Surface Condition

In this section, semiconductor surface characterization methods reviewed in Chapter 7 are assessed from the point of view of their usefulness in the in-line and on-line process monitoring in the manufacturing of semiconductor devices, as defined in Fig. 8.2. Only methods that do not require any additional treatment such as coating of the surface with thin film of material of any kind or thermal annealing of the sample implemented to obtain information regarding the condition of the surface or near-surface region are considered. In other words, in agreement with the scope of this book, the review is limited to the techniques compatible with the characterization of as-processed, bare semiconductor surfaces. The corresponding techniques are considered here, following the terminology adopted in Fig. 8.2.

**Analytical methods.** Among the methods in this group (see the discussion in Section 7.2), X-ray photoelectron spectroscopy (XPS), total reflection X-ray fluorescence (TXRF) spectroscopy, Auger electron spectroscopy (AES), and time-of-flight secondary ion mass spectroscopy (TOFSIMS) represent an integral component of the semiconductor surface research including process development and process diagnostics. Testifying to the long-established usefulness of analytical methods in semiconductor surface engineering, including process monitoring applications, the Auger spectroscopy, for instance, was used to study chemical transformations on silicon surfaces during thermal oxidation since an early stage of the MOSFET technology development (Ruzyllo *et al.*, 1976). The same applies to the XPS method with the results representing the chemical composition of the studied surface shown earlier in this volume.

All in all, the analytical methods identified above are indispensable in surface science and materials research, process development and diagnostics (Fig. 8.1), as well as in off-line process monitoring (Fig. 8.2). On the other hand, however, the methods considered are not compatible with the needs of real-time in-line process monitoring on product wafers.

To work around these limitations, the results of spectroscopic analysis of surface condition can be correlated, best at the process development stage, with the results obtained by applying other methods used in semiconductor surface characterization. Once adequately calibrated, the variability of surface characteristics that would be determined based on the use of analytical techniques can be detected using other methods, such as, for instance, measurements of wetting angle or electrical properties of silicon surface and near-surface region.

As an example of this approach, the results in Fig. 8.9 show changes in oxygen content on the silicon surface determined by XPS measurements and the wettability expressed in terms of the wetting (contact) angle on the same surface as a function of time of exposure to oxide etching in $HF:H_2O$ solution. As seen in this figure, the gradual decrease of the $O_{1s}$ peak on the Si surface is accompanied by changes in the value of the contact angle, indicating a transition from the hydrophilic to hydrophobic condition. Based on the results shown in Fig. 8.9, it can be concluded that when the residual native oxide etching is calibrated with respect to XPS results and then carried out in a controlled and reproducible fashion, measurements of the contact angle can be effectively used to monitor its performance (Ruzyllo *et al.*, 1996).

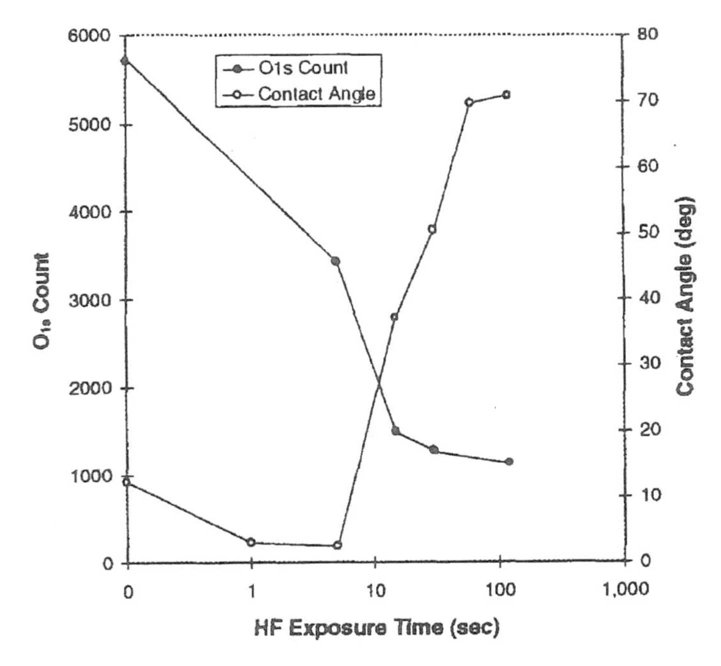


**Fig. 8.9.** Changes of the oxygen content determined by XPS expressed in terms of $O_{1s}$ count and related to the values of the contact (wetting) angle of the silicon surface as a function of time of exposure to $HF(1):H_2O(100)$ solution.

**Ellipsometry.** As discussed in Section 7.3, using ellipsometry or spectroscopic ellipsometry, surface properties can be obtained based on optical measurements but only in the case of the surface covered with a thin film of material at least few nanometers thick. The film should be transparent to the wavelengths used to allow the light to pass through. Furthermore, surfaces featuring excessive roughness may cause scattering of light away from the detector which would additionally limit the usefulness of ellipsometry in in-line process monitoring of the processes performed on the bare surfaces of semiconductor wafers.

The above features of ellipsometry render it compatible with in-line monitoring of the operations performed on the as-processed bare semiconductor surfaces only in the case of selected processes, particularly those in which process malfunction causes the spontaneous formation of the film-like deposit on the processed surface. Furthermore, ellipsometry can provide unique information regarding additive procedures during in-line monitoring of processes depositing thin films of material transparent to light on the monitored surface. For this reason, ellipsometers are an integral part of the semiconductor material characterization instrumentation typically available in the cleanrooms for on-line process monitoring, diagnostics, and development.

**Atomic Force Microscopy (AFM).** Considered earlier in this volume, AFM is the method designed to detect and quantify surface roughness with atomic-scale precision. For this reason, AFM tools are broadly used in semiconductor research laboratories and production facilities.

The AFM method is being implemented in various ways with static mode, also known as contact mode, being the most common (Fig. 7.11). Two other modes of AFM's operation, which fall into the dynamic mode category, are tapping mode and non-contact mode. In the case of this last mode, the AFM method can be considered as non-invasive. In addition, AFM characterization can be carried out in the air under atmospheric pressure or in a vacuum with the highest resolution AFM images of atomic structures obtained in an ultra-high vacuum environment. No special preparation of the surface prior to AFM characterization is needed, which makes this technique potentially compatible with the needs of in-line process monitoring in the case of the procedures in which malfunction would likely manifest itself in the roughening of the surface.

Several examples of AFM results were presented throughout this book, for instance, in Figs. 4.4 and 4.6 of Chapter 4 and Figs. 6.18 and 6.20 of Chapter 6.

**Wetting (contact) angle measurements.** Characterization of semiconductor surfaces aimed at the detection of variations in surface energy is based on the measurements of the wetting (contact) angle discussed in Section 7.5 and used in the experiment results of which are shown in Fig. 8.9. The method requires dispensing of the droplet of liquid on the studied surface and, thus, needs to be considered invasive. As such, it cannot be applied in the in-line process monitoring on product wafers but instead plays an important role in semiconductor device engineering in the on-line or off-line characterization of surface conditions.

As a further example of the relevance of contact angle measurements in the monitoring of the condition of semiconductor surfaces, Fig. 8.10 shows the changes in the surface recombination lifetime obtained from the SPV-based measurements and wetting angle measured on the same Si wafer before and after removal of the residual native oxide by means of AHF:methanol etching (Brubaker *et al.*, 1998).

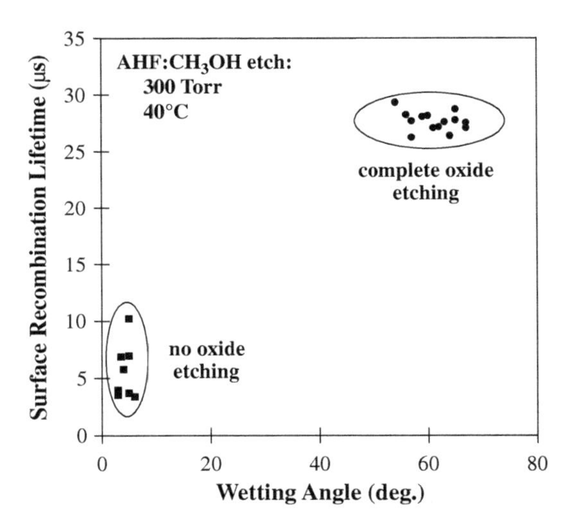


**Fig. 8.10.** Experimental results correlating surface recombination lifetime and wetting angle on silicon surface etched in anhydrous HF:methanol vapor.

**Electrical characterization of as-processed bare semiconductor surfaces.** Methods of electrical characterization of semiconductor surfaces are considered here in the context of the needs of in-line and on-line process monitoring on the as-processed wafers with exposed bare surfaces, in other words, on the wafers which were not subjected to any additional treatments prior to measurement.

Overall, electrical characterization of semiconductor surfaces is very well established in semiconductor device engineering and plays a pivotal role in the broad range of characterization applications. Over the years, the methods and methodologies of electrical characterization were subjected to changes in response to the introduction of new processing methods and the needs of varied configurations of surfaces subjected to monitoring. However, the general principles underlying the electrical characterization of semiconductor surfaces discussed remain unchanged.

As discussed in Chapter 7, electrical characterization of semiconductor surfaces can be implemented directly on the as-processed surface or can be based on the results of electrical characteristics of test devices such as MOS capacitor or Schottky diode formed on these surfaces. Since in this last case, the tested surface is altered by making the device, the methods of surface characterization relying on the measurements of test structures are not compatible with in-line process monitoring. This observation doesn't change the fact that the electrical characterization of semiconductor surfaces and interfaces using test structures in various configurations is an integral part of the process development methodology as well as diagnostics of possible process malfunctions in semiconductor device research and manufacturing.

To ensure the suitability of any given electrical method of semiconductor surface characterization for in-line, real-time process monitoring on product wafers, the method has to meet the following requirements. First, it should not alter in any way the condition of the surface, or in other words, it must be non-invasive. It means that it should not involve surface-altering physical contact between the monitored surface and the probe. Furthermore, it should not require alteration of its condition by illumination with light-carrying energy potentially affecting surface condition nor should it require thermal annealing or local changes of surface temperature.

The use of the term "non-invasive" in the description of in-line monitoring of semiconductor surfaces during device manufacturing warrants additional considerations. The point here is that the term "non-invasive" is not synonymous with the term "non-contact" understood as no physical contact to the surface.

In some surface characterization instruments referred to as non-contact, electrically neutral materials can be used to ensure a predetermined, reproducible distance between the measuring probe and the surface locally affecting its condition with a different effect in the case of different materials and surface features. Therefore, the extent of the impact of any given process monitoring technique on the condition of the surface needs to be determined for a specific surface and a specific process through additional dedicated experiments. In other words, the decision regarding the use of any given surface monitoring instrumentation should be preceded by a thorough investigation of interactions of the measurement stimulating energy with monitored surfaces and the potential impact of the measuring probe parts coming into contact with the measured surface.

## 8.6 Electrical Characterization Methods Proposed for In-line Monitoring and On-line Diagnostics of Semiconductor Surface Processing

In general, the advantage of electrical methods applied in semiconductor process monitoring is that they unravel surface and near-surface characteristics which are directly affecting the performance of the final device. While not qualitative, the electrical methods can provide in real time quantitative results indicating malfunction of the monitored process (Kamieniecki, Ruzyllo, 2001).

To exemplify the usefulness of electrical characterization in the semiconductor surface process monitoring applications, experimental results of SPV non-contact measurements of the density of surface charge and photoconductance decay (PCD) based near-surface characterization of semiconductor surfaces involving temporary lateral contact are considered in this section. As the earlier discussion indicated, these two methods could provide solutions to the challenges of in-line and on-line semiconductor process monitoring, respectively, and contribute to the broadly understood quality control in semiconductor device manufacturing.

The methods considered here perform different functions in the sequence concerned with surface characterization in semiconductor device manufacturing illustrated in Fig. 8.11. The SPV-based measurement of the density of surface charge $Q_s$ is a part of the in-line process monitoring component in the sequence. If the case of process malfunction manifesting itself by the significant departure of $Q_s$ from the assumed value is detected, the wafer is subjected to an online process diagnostics procedure based on the temporary contact PCD measurements. If the sequences presented in Fig. 8.11 were to be implemented in the cluster shown in Fig. 8.6, then the wafer indicating process malfunction would have to be removed from the tool and subjected to on-line process diagnostics applying temporary contact PCD measurements.

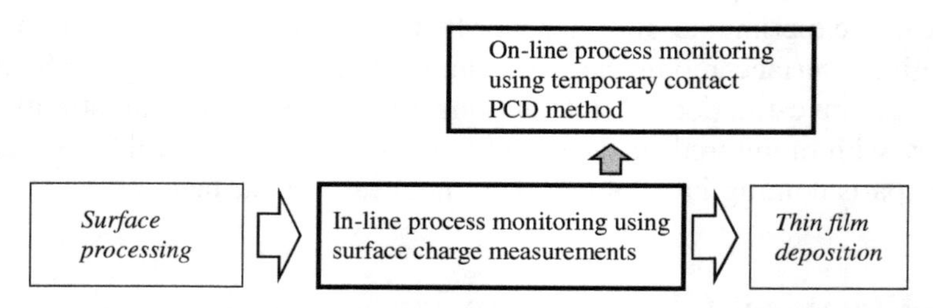


**Fig. 8.11.** Surface processing thin-film deposition sequence with suggested in-line process monitoring and on-line process diagnostics procedures.

The goal of PCD measurements is to determine whether process failure is related to the interactions at the surface of the wafer affecting the electronic properties of the near-surface region, specifically charge carrier transport characteristics. This last one is accomplished by measuring minority carrier lifetime and mobility of charge carriers in the near-surface region using temporary contact PCD characterization. Based on the results obtained, modifications of the monitored process, whether it is surface cleaning or etching, may be warranted.

What needs to be emphasized at this point is that the process monitoring-diagnostics scenario shown in Fig. 8.11 is a suggested solution addressing the control of the subtractive processes applied in FEOL semiconductor device technology. It is based on the author's research experiences and it is not the only approach addressing challenges at hand.

In the continuation of this discussion, experimental results showing the compatibility of the methods considered with the goals of in-line process monitoring and on-line process diagnostics are considered. The results included in this discussion are meant to demonstrate a correlation between the parameters measured using proposed techniques and surface characteristics obtained using other methods.

**Process monitoring based on surface charge measurements.** Among surface characteristics considered in Chapter 7, the density of surface charge *Qs* is a parameter dependent upon a broad range of surface features, both physical and chemical in nature. Therefore, non-contact methods determining the density of surface charge are considered suitable for in-line process monitoring. To reiterate points made earlier, the decision of whether in-line process monitoring can be performed using any given method on product wafers needs to be determined for each process monitoring application through the dedicated set of experiments.

As indicated in the discussion in Chapter 7, an advantageous characteristic of the measurements based on the determination of surface charge is the sensitivity of this parameter to essentially any change in surface characteristics (Kamieniecki *et al.*, 1994). Based on this feature, the non-contact technique considered here is seen as the one meeting requirements of in-line process monitoring, possibly on the product wafers, in the case where RF-enhanced coupling allowing separation between electrode and measured surface in the range of millimeters is implemented (Kamieniecki, 2011).

The high sensitivity of surface charge to the changes in the chemical makeup of the (100) surface of the same *p*-type silicon wafer exposed to various chemistries was demonstrated in Figs. 7.16 and 7.17. A consideration regarding surface charge measurements is that by applying procedures briefly described earlier in this volume, based on the discussed measurements of surface charge variations in the value of surface recombination lifetime can be linked to the changes of surface features measured using other techniques. For instance, Fig. 8.12 shows the correlation between surface recombination lifetime derived from surface charge measurement with XPS-determined $O_{1s}$ count representing oxide coverage of the surface as a function of time of Si surface exposure to the same HF(1):$H_2O$(100) solution (Roman *et al.*, 1998).

The results shown in the Fig. 8.12 further testify to the relevance of surface charge measurements in the representation of variations of the chemical composition of silicon surfaces and their impact on electronic characteristics of the near-surface region.

To ensure reliability of surface charge measurements in the monitoring of processes performed on Si surface, it is essential that the coverage of the surface with spontaneously grown native oxide remains unchanged from run to run. A study of surface charge evolution during the early stage of thermal oxidation of silicon demonstrated that even silicon oxide as

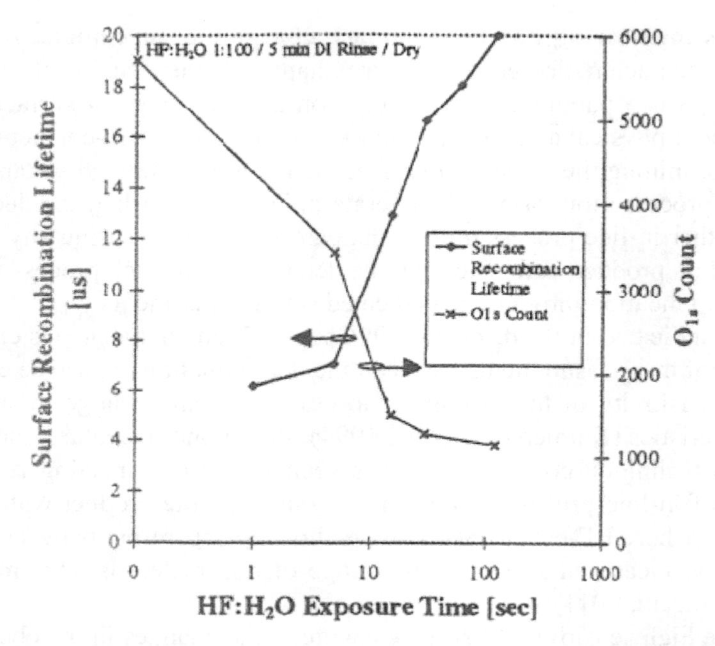


**Fig. 8.12.** Derived from SPV measured surface charge, surface recombination lifetime increases with decreasing oxygen count on the surface and resulting in increasing hydrogen termination of broken Si bonds on the surface.

thin as 2 nm brings about distinctly different surface charge $Q_s$ than the one measured under the exact same conditions on the same, but residual oxide-free, Si wafer.

Using the surface charge measurements methodology discussed in this review, variations of the surface charge density on Si surface subjected to thermal oxidation were investigated. As shown in Fig. 8.13, during the early stage of oxidation, up to the oxide thickness of about 3 nm, surface charge measured on *p*-type silicon surface goes through the progressive reduction of oxidation-related surface charge up to the oxide thickness of roughly 2 nm followed by the gradual increase in positive charge, reaching its final value at the oxide thickness of about 3–3.5 nm (Wang *et al.*, 2003). Changes in the density of surface charge observed are related to the changes in the chemical composition of Si surface during the early stage of thermal oxidation of silicon (Ruzyllo, 1978).

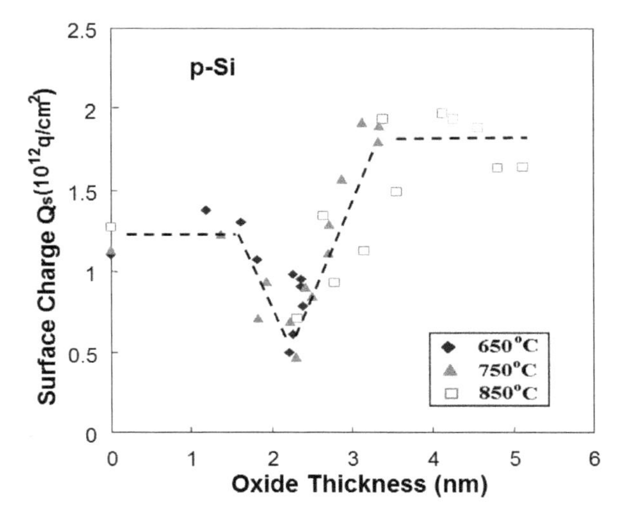


**Fig. 8.13.** Changes of the density of surface charge as a function of oxide thickness at three different temperatures of thermal oxidation of silicon in dry oxygen.

As demonstrated in the experimental results presented above, the measurements of the surface charge using a method which does not require electrical contact with the measured semiconductor surface provide information about its condition sufficient to monitor the performance of the processes to which bare semiconductor surface may be exposed during device manufacturing. To make it compatible with in-line process monitoring on product wafers, however, a non-invasive technique based on RF coupling should be considered (Kamieniecki, 2011).

**Process diagnostics based on temporary contact PCD measurements.** In terms of the process monitoring applications, it is expected that the results of in-line surface charge measurements remain unchanged from run to run. In such cases, no additional measurements of surface characteristics are needed and the wafer can be subjected to the next process in the device manufacturing sequence. If, however, run-to-run variations in the values of surface charge are detected, then according to the scenario presented in Fig. 8.11, additional measurements, this time on-line and if needed also off-line on the wafer removed from the production lot (Fig. 8.2), ought to be carried out to diagnose the cause of the process malfunction.

It is proposed in this discussion that diagnosis of the process malfunction is to be performed using temporary contact photoconductance decay (PCD) measurements in the lateral configuration (Figs. 7.14 and 7.19). This way, by investigating the electronic characteristics such as minority carrier lifetime in the near-surface region, insights into the nature of malfunction of the process can be obtained by isolating its effect on the components $Q_1$ and $Q_2$ comprising surface charge $Q_s$ (Fig. 4.7). Implementation of PCD measurements using temporary contacts by applying tungsten probes to the investigated surface through the top residual layer restricts the evaluation of phenomena taking place to the surface region (charge $Q_1$) with a marginal effect of chemical interactions taking place at the surface (charge $Q_2$).

Since the temporary contact PCD measurements are carried out using sharp-tipped probes, there is no need to apply any pre-measurement chemical treatments aimed at the removal from the surface of any residual film including spontaneously grown native oxide. In case the results of PCD characterization are not conclusive regarding the nature of the detected in-line process malfunction, the process diagnostics procedure needs to continue off-line. In off-line laboratory facilities, in which most of the methods discussed in Sections 6.3–6.6 of this *Guide* are commonly available, the exact nature of the process failure originally detected in-line based on the monitoring performed and the product wafers can be determined.

In support of the above-considered process malfunction diagnostics scenario, experimental results illustrating the usefulness of PCD measurements in this process are presented below. The results in Fig. 8.14 demonstrate the efficiency of PCD measurements in sensing the extent of surface damage as a function of the time of ion etching of silicon surface using chemically neutral $Ar^+$ ions (Arora *et al.*, 2016).

As seen in Fig. 8.14, PCD signal decay changes, depending on the etch time which when increased causes increased damage to the exposed silicon surface. Based on the shape of PCD plots and the rate of signal decay, the lifetime of minority carriers was calculated for various etch times (Drummond, Kshirsagar *et al.*, 2011). The results shown in Fig. 8.14 indicate the decrease in minority carrier lifetime affecting electrical conductivity of the near-surface region from 52 μs for the surface that was not exposed to high-energy $Ar^+$ ions to 6.35 μs for the surface subjected to interactions with high-energy $Ar^+$ ions for 2 minutes. This observation confirms the dependence of the minority carrier lifetime in the

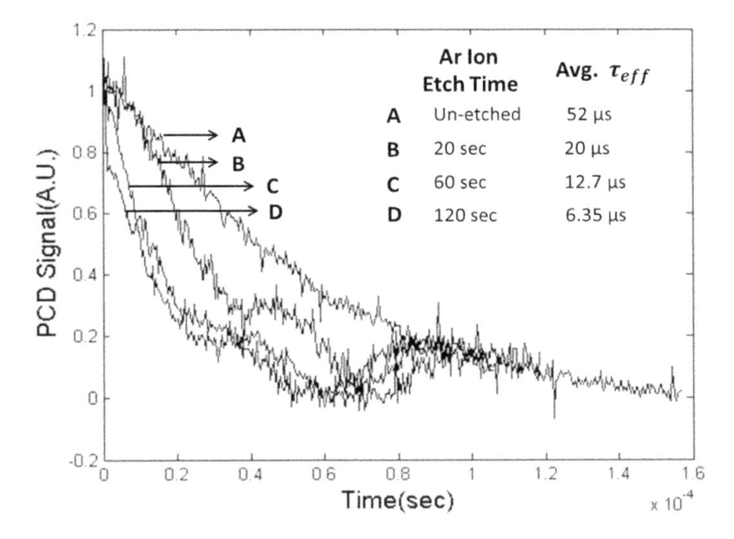


**Fig. 8.14.** Decrease of the minority carrier lifetime $\tau_{eff}$ derived from PCD measurements as a function of argon ion etching of Si surface.

near-surface region on the structural defectivity of this region, this time caused by surface exposure to heavy $Ar^+$ ions carrying significant kinetic energy.

To further underscore the relevance of the PCD measurements in the evaluation of the condition of semiconductor surfaces through its correlation with other surface characteristics, Fig. 8.15 shows the experimentally observed dependence of minority carrier lifetime measured using temporary contact and the thickness of thermally grown oxide remaining on Si surface between the probes in the areas from which thermal oxide was removed (Drummond, Bhatia *et al.*, 2011, 2013). The increase in the minority carrier lifetime shown in the figure results from the gradual increase in thickness of thermally grown $SiO_2$ penetrating deeper into the near-surface region of silicon wafer into the region featuring lesser distortion of the crystallographic structure. As a result, due to the reduced density of defects acting as recombination centers, an increase of the minority carrier lifetime is observed.

As indicated in the discussion in this section, there are methods of electrical characterization that allow the determination of the electronic characteristics of the semiconductor surface and sub-surface region without the processing of the test devices featuring permanent contacts to the

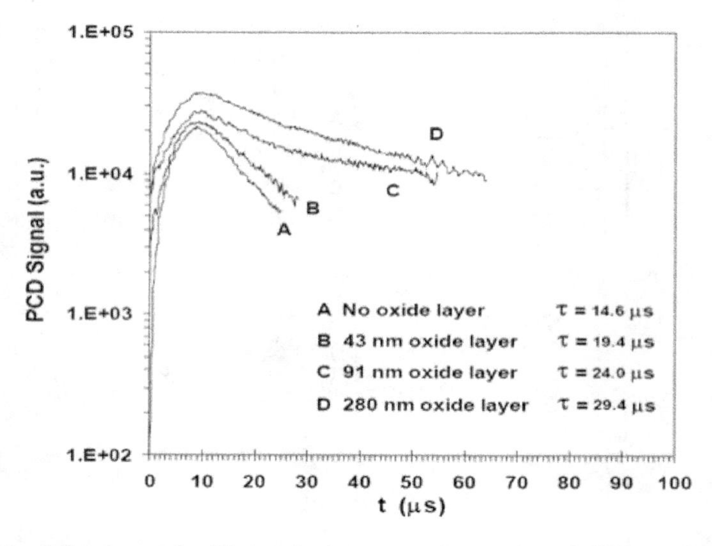


**Fig. 8.15.** Minority carrier lifetime in the near-surface region of silicon wafer derived based on PCD signal as a function of the thickness of thermally grown $SiO_2$.

measured surfaces. When based on RF coupling, the non-contact methods discussed above can be considered for in-line monitoring of surface processing steps carried out on product wafers.

A general point summarizing the discussion in this book is that surface characterization is an integral part of any semiconductor device engineering effort. The methods of semiconductor material and process characterization among those considered in Chapter 7, when selected appropriately and applied in combination, can both detect process malfunction and provide information regarding reasons for process failures typically encountered in semiconductor device manufacturing involving the processing of semiconductor surfaces.

A very final note is concerned with the fact that due to the changes in the architecture of semiconductor devices, both discrete and integrated, new materials used as well as new processing methods and tools introduced, points made in the book, while still valid, will evolve accordingly with time.

# Closing Remarks

As a vital part of semiconductor device engineering, and independently as a technical domain contributing to the progress of the semiconductor industry, semiconductor surface processing technology continues to evolve. As certain points made earlier in this volume need to be reinforced, the purpose of the brief closing remarks is to underscore features of semiconductor device engineering that affect semiconductor surface technology.

**Various device structures.** While there is a tendency to see semiconductor device structures in terms of specific applications, for instance, in advanced logic ICs, the reality is such that semiconductor devices are based on complex material systems featuring different configurations with varied surface topography depending on the device's function.

**3D surfaces.** The departure from surface planarity is an established trend which will continue in the future. As pointed out on several occasions in this volume, processed surfaces often depart from planarity as exemplified by the architecture of transistors used in logic ICs, including FinFETs and Gate All-Around (GAA) structures.

**Geometrical confinement.** Processing of the surfaces of nanoconfined materials, such single-atom-thick graphene, as well as nanosheets, nanowires, nanotubes, and nanodots does not involve surface treatments applied in the fabrication of devices using geometrically relaxed material systems.

**Structural fragility.** The fragile patterns created on the surface on a nanoscale call for solutions that prevent pattern collapse during surface processing.

**Quantum effects.** In the case of quantum effects driving device performance, the role of surface quality needs to be seen differently compared to devices operating on the rules of classical physics.

**Optical devices.** By virtue of photons being more efficient information carriers than electrons, it is a matter of developing suitable technologies before devices such as optical transistors will at least in some applications replace conventional electronic transistors. In such cases, interactions between surfaces and information carriers will be governed by different physical principles, and surface processing in device technology will need to be addressed differently compared to the fabrication of electronic devices.

**Surface accessibility.** There are semiconductor devices processed on the surfaces not directly exposed to process medium, such as in the case of devices featuring complex geometrical features, or surfaces created in the bulk of the material as it is a case in Micro-Electro-Mechanical Systems (MEMS). Dedicated process solutions are needed in such cases to ensure adequate performance of surface processing techniques and to allow control over the outcome of surface treatments.

**Surface area.** Smaller and more geometrically complex features on the one hand and large substrates processed in the case of solar cells on the other hand cause the implementation of surface processing steps to be dependent upon the area of processed surfaces.

**Semiconductor materials.** In addition to silicon, control over the surface characteristics of wide-bandgap compound semiconductors, gallium nitride (GaN) and silicon carbide (SiC), is also of interest. In the former case, the processing of the silicon surface prior to epitaxial deposition of the GaN layer is an added challenge. Additionally, with diamond being a semiconductor material suitable for devices operating at temperatures up to 800°C, processing of diamond surfaces brings about additional challenges. Furthermore, in thin-film semiconductor device technology including devices formed using organic semiconductors, different than in the case of devices formed on bulk substrates surface processing considerations apply.

**Substrates.** The shape and size of substrates evolving in some applications toward large, flexible, ribbon-shaped sheets call for dedicated solutions regarding surface processing. Also, non-semiconductor substrates, including sapphire, require established surface processing methods.

**Wearable electronics and photonics.** Wearable and implantable semiconductor electronic and photonic devices represent an important part of broadly understood semiconductor engineering. The way in which such devices are manufactured and installed calls for the consideration of surface processing steps from specific perspectives.

**Cleaning technology.** The processes used for semiconductor surface cleaning are also used for surface conditioning purposes. The focus on the prevention of contamination of the process and storage ambient in the course of a device's manufacturing further alters the role that surface cleaning technology plays.

**Surface engineering.** It involves local alterations of surface chemistry, promoting desired interactions in designated areas following patterns defined by surface engineering steps.

**Process integration.** Semiconductor manufacturing employing complex cluster tools integrated at several levels is a way to maintain processed surfaces in a controlled, protected environment.

**In-line process monitoring.** Control over the processes designed to alter in a predetermined fashion surface characteristics need to be monitored in-line, in real time on the production wafers. Non-contact, non-invasive electrical methods play a special role in this application, performing required measurements on the wafers in motion. Over time, robotics-based, AI-supported solutions are expected to facilitate the implementation of in-line process monitoring tasks.

# Bibliography


Allen, J. J. (2005). *Micro Electro Mechanical System Design*. CRC Press, Taylor and Francis.

Aspnes, D. S., and Handler, P. (1966). Surface states on cleaved (111) silicon surfaces. *Surface Science*, 4(4), 353–358.

Arora, A., Drummond, P. J., and Ruzyllo, J. (2016). Electrical characterization of silicon-on-insulator wafers by photoconductance decay (PCD) method. *ECS Journal of Solid-State Science and Technology*, 5(4), 3069–3074.

Baca, A. G., and Ashby, C. I. H. (2005). *Fabrication of GaAs Devices*. The Institution of Electrical Engineers.

Banerjee, S. (2015). Cryoaerosol cleaning of particles from surfaces. In *Particles Adhesion and Removal*. Wiley.

Beck, S. E., Bohling, D. A., Felker, B. S., George, M. A., Gilicinski, A. G., Ivankovits, J. C., Langan, J. G., Rynders, S. W., Norman, J. A. T., Roberts, D. A., Voloshin, G., Hess, D. M., and Lane, A. (1994). Chemical vapor cleaning technologies for dry processing in semiconductor manufacturing. *Electrochemical Society Proceedings*, 94-7, 253–267.

Benedek, G., and Toennis, P. (2018). *Atomic Scale Dynamics at Surfaces*. Springer.

Brubaker, M., Roman, P., Staffa, J., and Ruzyllo, J. (1998). Monitoring of chemical oxide removal from silicon surfaces using a surface photovoltage technique. *Electrochemical and Solid State Letters*, 1, 130–132.

Butterbaugh, J., and Muscat, A. (2007). Gas-phase wafer cleaning technology. In K. A. Reinhardt and W. Kern (Eds.), *Handbook of Semiconductor Wafer Cleaning Technology*. William Andrew.

Caymax, M., Decoutere, S., Rohr, E., Wandervorst, W., Heyns, M., Sprey, H., Storm, A., and Maes, J.-W. (1998). Electrical evaluation of the epi/substrate interface quality after different in-situ and ex-situ low-temperature pre-epi cleaning methods. In *Ultra-Clean Processing of Silicon Surfaces*. Scitec Publications.

Chao, J-H., Kshirsagar, A., and Ruzyllo, J. (2013). Surface processing for area selective mist deposition of nanocrystalline quantum dot films. *ECS Transactions*, 58(6), 311–316.

Chang, K., Shanmugasundaram, K., Lee, D-O., Roman, P., Wu, C.-T., Wang, J., Shallenberger, J., Mumbauer, P., Grant, R., Ridley, R., Dolny, G., and Ruzyllo, J. (2004). Silicon surface treatments in advanced MOS gate processing. *Microelectronic Engineering*, 72, 130–135.

Chang, K., Witt, T., Hoff, A., Woodin, R., Ridley, R., Dolny, G., Shanmugasundaram, K., Oborina, E., and Ruzyllo, J. (2005). Surface roughness in silicon carbide technology. In *Cleaning Technology in Semiconductor Device Manufacturing IX, ECS Transactions*, 1(3), 228–235.

Chin, A., Chen, W. J., Chang, T., Kao, R. H., Lin, B. C., Tsai, C., and Huang, J. C. M. (1997). Thin oxides with in-situ native oxide removal. *IEEE Electron Device Letters*, 18, 417–419.

Chongsawangvirod, S., and Irene, E. A. (1991). A spectroscopic differential reflectometry study of (100), (110), (111), (311), and (511) silicon surfaces. *Journal of the Electrochemical Society*, 138(6), 1748–1752.

Chu, R. (2021). GaN MOS structures with low interface trap density. *Solid State Phenomena*, 314, 79–83.

Dabrowski, J., and Mussig, H.-J. (2000). *Silicon Surfaces and Formation of Interfaces: Basic Science*. World Scientific Publishing Co.

Daffron, C., Torek, K., Ruzyllo, J., and Kamieniecki, E. (1994). Removal of Al from silicon surfaces using UV/$Cl_2$. *Electrochemical Society Proceedings*, 94-7, 281–288.

Danel, A., Tsai, C.-L., Shanmugasundaram, K., Tardif, F., Kamieniecki, E., and Ruzyllo, J. (2003). Cleaning of Si surfaces by lamp illumination. *Solid State Phenomena*, 92, 195–198.

Deal, B. E., McNeilly, M. A., Kao, D. B., and deLarios, J. M. (1990). Vapor phase wafer cleaning and integrated processing. *Electrochemical Society Proceedings*, 90-9, 121–128.

Decker, E. L., and Garoff, S. (1999). Physics of contact angle measurement. *Colloids and Surfaces A: Physicochemical and Engineering Aspects*, 156(1-3), 177–189.

Drummond, P., Ramani, K., and Ruzyllo, J. (2009). Monitoring of semiconductor surfaces using photoconductance decay (PCD) method. *ECS Transactions*, 25(5), 361–366.

Drummond, P. J., Bhatia, D., Kshirsagar, A., Ramani, S., and Ruzyllo, J. (2011). Studies of photoconductance decay method for characterization of near-surface electrical properties of semiconductors. *Thin Solid Films*, 519, 7621–7626.

Drummond, P., Kshirsagar, A., and Ruzyllo, J. (2011). Characterization of near-surface electrical properties of multi-crystalline silicon wafers. *Solid-State Electronics*, 55, 29–33.

Drummond, P., Bhatia, D., and Ruzyllo, J. (2013). Measurement of effective carrier lifetime at the semiconductor–dielectric interface by photoconductive decay (PCD) method. *Solid-State Electronics*, 81, 130–135.

Drummond, P. J., Wali, A., Barth, M. J., Diehm, A. M., Datta, S., and Ruzyllo, J. (2017). Photoconductance decay characterization of 3D multi-fin silicon on SOI substrates. *IEEE Electron Device Letters*, 38(11), 1513–1516.

Erdamar, M., Roman, P., Mumbauer, P., Klimkiewicz, M., and Ruzyllo, J. (2008). Deep lateral anhydrous etching for MEMS release processes. *Journal of Micro-Nanolithography MEMS and MOEMS*, 7(3), 1–8.

Fonash, S. (1990). An overview of dry etching damage and contamination. *Journal of the Electrochemical Society*, 137, 12–15.

Frystak, D. C., and Ruzyllo, J. (1992). Pre-oxidation cleaning of silicon using remote plasma. *Electrochemical Society Proceedings*, 92-12, 58–64.

Fujiwara, H. (2007). *Spectroscopic Ellipsometry, Principles and Applications*. John Wiley and Sons.

Gosh, A., Subrahmanyam, K. S., Krishna, K. S., Datta, S., Govindaraj, A., Pati, S. K., and Rao, C. N. R. (2008). Uptake of $H_2$ and $CO_2$ by graphene. *Journal of Physical Chemistry C*, 112(40), 15704–15707.

Grebs, T., Ridley, R., Chang, K., Wu, C.-T., Agarwal, R., Mytych, J., Dimachkie, W., Dolny, G., Michalowicz G., and Ruzyllo, J. (2004). Effect of RIE sequence and post-RIE surface processing on the reliability of gate oxide in a trench. *Electrochemical Society Proceedings*, 2003-26, 348–355.

Grove, A. S. (1967). *Physics and Technology of Semiconductor Devices*. John Wiley & Sons.

Gu, T., Ditizio, R. A., Fonash, S. J., Awadelkarim, O. O., Ruzyllo, J., and Collins, R. W. (1994). Damage to Si substrates during $SiO_2$ etching: A comparison of reactive ion etching and magnetron enhanced reactive ion etching. *Journal of Vacuum Science & Technology B*, 12(2), 567–572.

Hattori, T. (1998). *Ultraclean Surface Processing of Silicon Wafers*. Springer-Verlag.

Haugstad, G. (2012). *Atomic Force Microscopy: Understanding Basic Models and Advanced Applications*. Wiley.

Heyns, M., Mertens, P., and Ruzyllo, J. (1999). Advanced wet and dry cleaning coming together for next generation. *Solid State Technology*, 42(3), 37–42.

Hoff, A. M., and Ruzyllo, J. (1988). Atomic oxygen and thermal oxidation of silicon. *Applied Physics Letters*, 52(15), 1264–1266.

Hossain, S., Pantano, C., and Ruzyllo, J. (1990). Removal of surface organic contaminants during thermal oxidation of silicon. *Journal of the Electrochemical Society*, 137(10), 3287–3290.

Hwang, D. K., Ruzyllo, J., and Kamieniecki, E. (1994). Cleaning of silicon surfaces after RIE using UV/ozone and HF/$CH_3OH$. *Electrochemical Society Proceedings*, 94-7, 401–407.

Hwang, D. K., Torek, K., and Ruzyllo, J. (1996). Slight etching of silicon to control post-RIE damage. *Proceedings of First International Symposium on Plasma Process Induced Damage*, 137–142.

Hwang, E., Eaton, C., Mujumdar, S., Madan, H., Ali, A., Bhatia, D., Datta, S., and Ruzyllo, J. (2011). Processing and characterization of GaSb/high-k dielectric interfaces. *ECS Transactions*, 41(5), 157–162.

Irene, E. A. (2008). *Surfaces, Interfaces and Thin-Films for Microelectronics*. Wiley-Interscience.

Ishizaka, A., and Sharaki, Y. (1986). Low temperature surface cleaning of silicon and its application to silicon MBE. *Journal of the Electrochemical Society*, 133, 666–671.

Ito, T., Sugino, R., Watanabe, S., Nara, Y., and Sato, Y. (1990). UV-enhanced dry cleaning of silicon wafers. *Electrochemical Society Proceedings*, 90-9, 114–120.

Kamieniecki, E., Roman, P., Hwang, D., and Ruzyllo, J. (1994). A new method for in-line, real-time monitoring of wafer cleaning operations. *Acco Leuven*, 189–195.

Kamieniecki, E. (2001). Apparatus and method for rapid photo-thermal surface treatment [U.S. Patent No. 6,325,078 B2]. U.S. Patent and Trademark Office.

Kamieniecki, E., and Ruzyllo, J. (2001). Method for real-time in-line testing of semiconductors wafers [U.S. Patent No. 6,315,574 B1]. U.S. Patent and Trademark Office.

Kamieniecki, E. (2011). Electrical characterization of semiconductor materials [U.S. Patent No. 7,898,280 B2]. U.S. Patent and Trademark Office.

Kern, W., and Puotinen, D. (1970). Cleaning solutions based on hydrogen peroxide for use in silicon semiconductor technology. *RCA Review*, 31, 187–192.

Kern, W. (1990). The evolution of silicon wafer cleaning technology. *Journal of the Electrochemical Society*, 137(6), 1887–1892.

Kern, W. (Ed.). (1993). *Handbook on Semiconductor Wafer Cleaning Technology*. Noyes Publications.

Kingston, R. H. (Ed.). (2016). *Semiconductor Surface Physics*. University of Pennsylvania Press, Inc.

Kim, M. S., Lee, J. H., and Kwak, M. K. (2020). Review: Surface texturing methods for solar cell efficiency enhancement. *International Journal of Precision Engineering and Manufacturing*, 21, 1389–1405.

Kirby, K., Shanmugasundaraman, K., and Ruzyllo, J. (2007). Interactions of sapphire surfaces with standard cleaning solutions. *ECS Transactions*, 11(2), 343–348.

Klokenkamper, R., and Von Bohlen, A. (2015). *Total-Reflection X-Ray Fluorescence Analysis and Related Methods*. John Wiley and Sons.

Kurjata-Pfitzner, E. (1980). On the effect of surface recombination on some properties of semiconductor devices. *Surface Science*, 10(4), 259–264.

Kwok, D. Y., and Neuman, A. W. (1999). Contact angle measurement and contact angle interpretation. *Advances in Colloid and Interface Science*, 81(3), 167–174.

Lagowski, J., Kontkiewicz, A. M., Jastrzebski, L., and Edleman, P. (1993). Method for the measurement of long minority carrier diffusion lengths exceeding wafer thickness. *Applied Physics Letters*, 63(21), 2902–2904.

Lander, T. J., and Morrison, J. (1962). Low energy electron diffraction study of silicon surface structures. *Journal of Applied Physics*, 33, 2089–2092.

Law, A., Jones, L. O., and Walls, J. M. (2023). The performance and durability of anti-reflection coatings for solar module cover glass — A review. *Solar Energy*, 261(7), 85–92.

Le Roux, V., Machicoane, G., Kerdiles, S., Laffitte, R., Béchu, N., Vallier, L., Borsoni, G., Korwin-Pawlowski, M. L., Roman, P., Wu, C.-T., and Ruzyllo, J. (2004). Etching of silicon native oxide using ultraslow multicharged $Ar^{q+}$ ions. *Journal of the Electrochemical Society*, 151(1), G76–G81.

Lee, D-O., Roman, P., Wu, C.-T., Mumbauer, P., Brubaker, M., Grant, R., and Ruzyllo, J. (2002). Mist deposited high-k dielectrics for next generation MOS gates. *Solid-State Electronics*, 46, 1671–1677.

Lee, S. J., Imen, K., and Allen, S. D. (1993). Laser-assisted particle removal from silicon surfaces. *Microelectronic Engineering*, 20(1-2), 145–152.

Lindon, J. C., Tranter, G. E., and Koppenaal, D. W. (Eds.). (2017). *Encyclopedia of Spectroscopy and Spectrometry*. Elsevier Ltd.

Lukasiak, L., Roman, P., Jakubowski, A., and Ruzyllo, J. (2001). Analysis of surface and interface charge interactions in SOI substrates. *Solid-State Electronics*, 45, 95–99.

Lüth, H. (2015). *Solid Surfaces, Interfaces and Thin Films*. Springer-Verlag.

Luther, B. P., Ruzyllo, J., and Miller, D. I. (1993). Nearly isotropic etching of 6H SiC in $NF_3$ and $O_2$ using a remote plasma. *Applied Physics Letters*, 63(2), 171–173.

Ma, Y., Green, M. L., Torek, K., Ruzyllo, J., Opila, R., Konstandinidis, K., Siconolfi, D., and Brasen, D. (1995). In situ vapor phase pre-gate oxide cleaning and its effects on metal-oxide-semiconductor device characteristics. *Journal of the Electrochemical Society*, 142, L217–L219.

Maly, W., Singh, N. Z., Shen, N., Li, X., Pfitzner, A., Kasprowicz, D., Kuzmicz, W., Lin, Y-W., and Marek-Sadowska, M. (2011). Twin-gate, vertical slit FET (VESFET) for highly periodic layout and 3D integration. *International Conference on Mixed Design of Integrated Circuits and Systems — MIXDES 2011*, 145–150.

Marinskiy, D., Edelman, P., Lagowski, J., Loy, T. C., Almeida, C., and Savtchouk, A. (2016). Kelvin force microscopy and corona charging for semiconductor material and device characterization. *Superlattices and Microstructures*, 99, 13–17.

McGuire, G. E. (Ed.). (1989). *Characterization of Semiconductor Materials*. Noyes Publications.

Moghadam, F. K., and Mu, X.-C. (2002). A study of contamination and damage on Si surfaces induced by dry etching. *IEEE Transactions on Electronic Devices*, 36(9), 1602–1608.

Mönch, W. (2001). *Semiconductor Surfaces and Interfaces*. Springer.

Nishimura, J. (1991). Influence of thermally created sulfur vacancies on sublimation of cadmium sulfide crystals. *Japanese Journal of Applied Physics*, 30(3R), 537–541.

Ohira, F., and Itakura, M. (1979). Characterization of Si surface by ellipsometry. *Japanese Journal of Applied Physics*, 18(7), 1243–1248.

Ohmi, T., Kotani, K., Teramoto, A., and Miyashita, M. (1991). Dependence of electron mobility on Si-$SiO_2$ interface microroughness. *IEEE Electron Device Letters*, 12(12), 652–654.

Ohring, M. (2002). *Materials Science of Thin Films: Deposition and Structure*. Academic Press.

Philipossian, A., and Mustapha, L. (2003). Tribological characterization of post-CMP brush scrubbing. *Solid State Phenomena*, 92, 275–278.

Pierret, R. F. (1983). *Field Effective Devices*. Addison-Wesley.

Pierret, R. F. (1996). *Semiconductor Device Fundamentals*. Addison-Wesley.

Reinehardt, K. A., and Reidy, R. F. (Eds.). (2011). *Handbook of Cleaning for Semiconductor Manufacturing*. Wiley-Scivener.

Ridley, R., Wu, C-T., Roman, P., Dolny, G., Grebs, T., Stensney, F., and Ruzyllo, J. (2000). Thickness dependent sensitivity of gate oxides to surface contamination. *Electrochemical Society Proceedings*, 99-36, 158–163.

Roman, P., Hwang, D., Torek, K., Ruzyllo, J., and Kamieniecki, E. (1995). Monitoring of HF/$H_2O$ treated silicon surfaces using non-contact surface charge measurements. *Materials Research Society Proceedings*, 386, 401–406.

Roman, P., Staffa, J., Kamieniecki, E., and Ruzyllo, J. (1998). Surface dopant concentration monitoring using non-contact surface charge profiling. *Journal of Applied Physics*, 83(4), 2297–2301.

Roman, P., Brubaker, M., Staffa, J., Kamieniecki, E., and Ruzyllo, J. (1998). Non-contact monitoring of electrical characteristics of silicon surfaces and near-surface region. *American Institute of Physics Conference Proceedings*, 449, 250–255.

Roman, P., Lee, D.-O., Wang, J., Wu, C.-T., Subramanian, V., Brubaker, M., Mumbauer, P., Grant, R., and Ruzyllo, J. (2001). Gas-phase surface conditioning in a high-*k* gate stack cluster. *Electrochemical Society Proceedings*, 2001-26, 241–246.

Roman, P., Torek, K., Shanmugasundaraman, K., Mumbauer, P., Vestyck, D., Hammond, P., and Ruzyllo, J. (2009). Current advances in AHF/organic solvent processing of semiconductor surfaces. *Solid State Phenomena*, 145-146, 231–236.

Ruzyllo, J., Shiota, I., Miyamoto, N., and Nishizawa, J. (1976). Low temperature oxidation of silicon studied by photosensitive ESR and Auger electron spectroscopy. *Journal of the Electrochemical Society*, 123(1), 26–31.

Ruzyllo, J., Jakubowski, A., and Swit, A. (1978). Studies of surface conduction mechanism in silicon covered with ultrathin oxide film. *Bulletin of the Polish Academy of Sciences*, 26, 593–598.

Ruzyllo, J. (1978). Model of structural transformation on the chemically etched silicon surface during thermal oxidation. *Physica Status Solidi (a)*, 48, 199–204.

Ruzyllo, J. (1980). Lateral MIS tunnel transistor. *IEEE Electron Device Letters*, EDL-1(10), 197–199.

Ruzyllo, J., Duranko, G. T., and Hoff, A. M. (1987). Pre-oxidation UV treatment of silicon wafers. *Journal of the Electrochemical Society*, 134, 2052–2056.

Ruzyllo, J. (1988). Evaluating the feasibility of dry cleaning of silicon wafers. *Microcontamination*, 60(3), 39–43.

Ruzyllo, J., Hoff, A., Frystak, D., and Hossain, S. (1989). Electrical evaluation of wet and dry cleaning procedures for silicon device fabrication. *Journal of the Electrochemical Society*, 136, 1474–1478.

Ruzyllo, J. (1990). Issues in dry cleaning of silicon wafers. *Solid State Technology*, 33(3), S1–S5.

Ruzyllo, J., Frystak, D., and Bowling, R. (1990). Dry cleaning procedure for silicon IC fabrication. *IEEE International Electron Devices Meeting Technical Digest*, 409–412.

Ruzyllo, J. (1993). Overview of dry wafer cleaning processes. In W. Kern (Ed.), *Handbook of Semiconductor Wafer Cleaning Technology* (Ch. 3). Noyes Publications.

Ruzyllo, J., Torek, K., Daffron, C., Grant, R., and Novak, R. (1993). Etching of thermal oxides in low-pressure anhydrous HF/$CH_3OH$ gas mixture at elevated temperature. *Journal of the Electrochemical Society*, 140(4), L64–L66.

Ruzyllo, J., Roman, P., Staffa, J., Kashkoush, I., and Kamieniecki, E. (1996). Process monitoring using surface charge profiling (SCP) method. *Proceedings of SPIE*, 2876, 162–167.

Ruzyllo, J., Rohr, E., Caymax, N., Baeyens, M., Conard, T., Mertens, P., and Heyns, M. (1998). Role of UV/chlorine exposure during dry surface conditioning before integrated epi deposition process. *Scitec Publications*, 233–238.

Ruzyllo, J., Rohr, E., Baeyens, M., Mertens, P., and Heyns, M. (1999). Fluorine in thermal oxides from HF peroxidation surface treatments. *Electrochemical and Solid State Letters*, 2, 336–339.

Ruzyllo, J., and Roman, P. (1999). Electrical characterization of c-Si surfaces. In *Properties of Crystalline Silicon, EMIS Data Reviews Series* (No. 20). INSPEC Publications.

Ruzyllo, J. (2007). Semiconductor surface cleaning and conditioning challenges beyond planar silicon technology. *ECS Transactions*, 9(1), 87–92.

Ruzyllo, J. (2010). Semiconductor cleaning technology: Forty years in the making. *Electrochemical Society Interface*, 19(1), 44–49.

Ruzyllo, J. (2014). Assessment of the progress in gas-phase cleaning of silicon surfaces. *ECS Journal of Solid State Science and Technology*, 3(1), N3060–N3064.

Ruzyllo, J., and Drummond, P. J. (2016). Electrical characterization of as-processed semiconductor surfaces. *Solid State Phenomena*, 255, 299–304.

Ruzyllo, J. (2017). *Semiconductor Glossary*. World Scientific Publishing Co.

Ruzyllo, J. (2020). *Guide to Semiconductor Engineering*. World Scientific Publishing Co.

Sabeeh, A., Brigeman, A. N., and Ruzyllo, J. (2019). Performance of single-crystal silicon solar cells with mist-deposited nanocrystalline quantum dot downshifting films. *IEEE Journal of Photovoltaics*, 9(4), 1006–1012.

Saga, K., and Hattori, T. (1996). Identification and removal of trace organic contamination on silicon wafers stored in plastic boxes. *Journal of the Electrochemical Society*, 143(10), 3279–3283.

Sahari, S. K., Sing, J. C. H., and Hamid, K. A. (2009). The effects of RCA clean variables on particle removal efficiency. *International Journal of Electrical and Computer Engineering*, 3(2), 262–267.

Saito, H., Munakata, A., Ichishima, D., Yamanishi, T., Okamoto, A., Saga, K., Kuniyasu, H., and Hattori, T. (2004). Cleaning of fragile fine structures with cryogenic nitrogen aerosols. *Electrochemical Society Proceedings*, 2003-26, 289–294.

Saleh, B. E. A., and Teich, M. C. (1991). *Fundamentals of Photonics*. Wiley Interscience.

Sanders, W. (2019). *Atomic Force Microscopy, Fundamental Concepts and Laboratory Investigations*. CRC Press.

Schroder, D. K. (2015). *Semiconductor Material and Device Characterization*. John Wiley and Sons.

Shanmugasundaram, K., Chang, K., Shallenberger, J., Danel, A., Tardif, F., Meillerot, M., and Ruzyllo, J. (2004). Reversing of silicon surface aging by lamp cleaning. *Electrochemical Society Proceedings*, 2003-26, 108–113.

Shanmugasundaram, K., Chang, K., and Ruzyllo, J. (2005). Effect of silicon surface conditioning on film formation using mist deposition. *ECS Transactions*, 1(3), 105–110.

Staffa, J., and Ruzyllo, J. (1996). The effect of pre-etch surface hydration on the $SiO_2$ gas-phase etch process. *Symposium on Ultraclean Processing of Semiconductor Surfaces*, 261–266.

Staffa, J., Fakhouri, S., Brubaker, M., Roman, P., and Ruzyllo, J. (1999). Effects controlling initiation and termination of gas-phase cleaning reactions. *Journal of the Electrochemical Society*, 146, 321–325.

Suliman, S. A., Gallogunta, N., Trabzon, L., Hao, J., Dolny, G., Ridley, R., Benjamin, J., Kocon, C., Zeng, J., Knoedler, C. M., Horn, M., Awadelkarim, O. O., Fonash, S. J., and Ruzyllo, J. (2001). The impact of trench geometry and processing on the performance and reliability of low voltage power UMOSFETs. *Proceedings of IEEE International Reliability Physics Symposium*, 308–312.

Sung, P.-J., Chang, S.-W., Kao, K.-H., Wu, C.-T., Su, C.-J., Cho, T.-C., Hsueh, F.-K., and Lee, W.-H. (2020). Fabrication of vertically stacked nanosheet junctionless field-effect transistors and applications for the CMOS and CFET inverters. *IEEE Transactions on Electronic Devices*, 67(9), 3504–3509.

Symposia. The Electrochemical Society. (1989–2025). *Symposia on Surface Cleaning Science and Technology (SCST)*.

Symposia. Interuniversity Microelectronics Center IMEC vzv. (1992–2025). *Symposia on Ultraclean Processing of Semiconductor Surfaces (UCPSS)*.

Symposia. Linx Consulting. (2002–2025). *Surface Preparation and Cleaning Conference (SPCC)*.

Sze, S. M., and Ng, K. K. (2006). *Physics of Semiconductor Devices* (3rd ed.). Wiley-Interscience.

Tao, F., and Bernasek, S. L. (Eds.). (2012). *Functionalization of Semiconductor Surfaces*. Wiley.

Tompkins, H. G., and Irene, E. A. (Eds.). (2005). *Handbook of Ellipsometry*. William Andrew, Inc.

Torek, K., Mieckowski, A., and Ruzyllo, J. (1995). Evolution of Si surfaces after anhydrous HF/methanol etching. *Electrochemical Society Proceedings*, 95-20, 208–213.

Torek, K., Ruzyllo, J., Grant, R., and Novak, R. (1995). Reduced pressure etching of thermal oxides in anhydrous HF/alcoholic gas mixture. *Journal of the Electrochemical Society*, 142(4), 1322–1326.

Tsai, C.-L., Roman, P., Wu, C.-T., Pantano, C., Berry, J., Kamieniecki, E., and Ruzyllo, J. (2003). Control of organic contamination of silicon surfaces using white light illumination in ambient air. *Journal of the Electrochemical Society*, 150(1), G39–G43.

Wang, J., Roman, P., Kamieniecki, E., and Ruzyllo, J. (2003). Surface charge evolution during early stage of thermal oxidation of silicon. *Electrochemical and Solid-State Letters*, 6(5), G63–G65.

Wu, C.-T., Ridley, R., Roman, P., Dolny, G., Grebs, T., Hao, J., and Ruzyllo, J. (2001). The effect of surface treatments and growth conditions on electrical characteristics of thick (>50 nm) gate oxides. *Journal of the Electrochemical Society*, 148, F184–F188.

Wu, C.-T., Ridley, R. S., Dolny, G., Grebs, T., Knoedler, C., Suliman, S., Venkataraman, B., Awadelkarim, O., and Ruzyllo, J. (2002). Growth and reliability of thick gate oxide in U-trench for power MOSFET's. *Proceedings of 14th International Symposium on Power Semiconductor Devices and ICs*, 149–153.

Yew, T. R., and Reif, R. (1990). Low-temperature in situ surface cleaning of oxide-patterned wafers by Ar/$H_2$ plasma sputter. *Journal of Applied Physics*, 68, 4681–4685.

Yuh, H. K., Park, J. W., Hwang, K. H., Yoon, E., and Whang, K. W. (1998). Hydrogen plasma cleaning of oxide patterned Si wafers for low temperature silicon epitaxy. *Electrochemical Society Proceedings*, 97-35, 307–312.

Zhang, Z., and Yates, J. T. (2012). Band bending in semiconductors: Chemical and physical consequences at surfaces and interfaces. *Chemical Reviews*, 112(10), 5520–5551.

# Index

**M**



**N**

**Q**



**R**



**S**

www.ingramcontent.com/pod-product-compliance
Lightning Source LLC
LaVergne TN
LVHW012321090525
810667LV00002B/16

* 9 7 8 9 8 1 1 2 5 4 8 1 9 *